Advanced Verification Topics

Bishnupriya Bhattacharya
John Decker
Gary Hall
Nick Heaton
Yaron Kashai
Neyaz Khan
Zeev Kirshenbaum
Efrat Shneydor

©2011 Cadence Design Systems, Inc. All rights reserved worldwide. Published 2011.

Printed in the United States of America.

Cadence Design Systems, Inc. (Cadence), 2655 Seely Ave., San Jose, CA 95134, USA.

Open SystemC, Open SystemC Initiative, OSCI, SystemC, and SystemC Initiative are trademarks or registered trademarks of Open SystemC Initiative, Inc. in the United States and other countries and are used with permission.

Trademarks: Trademarks and service marks of Cadence Design Systems, Inc. contained in this document are attributed to Cadence with the appropriate symbol. For queries regarding Cadence's trademarks, contact the corporate legal department at the address shown above or call 800.862.4522. All other trademarks are the property of their respective holders.

Restricted Permission: This publication is protected by copyright law and international treaties and contains trade secrets and proprietary information owned by Cadence. Unauthorized reproduction or distribution of this publication, or any portion of it, may result in civil and criminal penalties. Except as specified in this permission statement, this publication may not be copied, reproduced, modified, published, uploaded, posted, transmitted, or distributed in any way, without prior written permission from Cadence.

The information contained in this document cannot be used in the development of like products or software, whether for internal or external use, and shall not be used for the benefit of any other party, whether or not for consideration.

Disclaimer: Information in this publication is subject to change without notice and does not represent a commitment on the part of Cadence. Except as may be explicitly set forth in such agreement, Cadence does not make, and expressly disclaims, any representations or warranties as to the completeness, accuracy or usefulness of the information contained in this document. Cadence does not warrant that use of such information will not infringe any third party rights, nor does Cadence assume any liability for damages or costs of any kind that may result from use of such information.

Restricted Rights: Use, duplication, or disclosure by the Government is subject to restrictions as set forth in FAR52.227-14 and DFAR252.227-7013 et seq. or its successor.

ISBN 978-1-105-11375-8

Contents

Preface		**xv**
1	**Introduction to Metric-Driven Verification**	**1**
	1.1 Introduction	1
	1.2 Failing to Plan = Planning to Fail	2
	1.3 Metric-Driven Verification	4
	1.4 Building Strong Testbench Foundations	5
	1.5 Simulation Isn't the Only Way	7
	1.6 Low Power isn't Just the Designer's Problem	8
	1.7 Reuse Isn't Just About Testbench Components	9
	1.8 Does Speed Matter?	10
	1.9 What About Scalability?	11
	1.10 Is Metric-Driven Verification Just for RTL Hardware?	12
	1.11 How Do I Get Up to Speed with All this New Stuff?	14
	1.12 Summary	14
2	**UVM and Metric-Driven Verification for Mixed-Signal**	**15**
	2.1 Why Metric-Driven Verification for Analog?	15
	2.2 Approach and Scope	16
	2.3 Planning for Analog Verification	20
	2.3.1 Including Analog Properties	20
	2.3.2 Verification Plan Structured for Reuse	22
	2.4 Constructing a UVM-MS Verification Environment	23
	2.4.1 Analog Verification Blocks	23
	2.4.2 Analog Configuration Interface	24
	2.4.3 Threshold Crossing Monitor	24
	2.5 Architecture of a Sample Testbench	25

2.6	Collecting Coverage		26
	2.6.1	Direct and Computed Coverage Collection	27
	2.6.2	Deciding on Coverage Ranges	30
	2.6.3	Trading Off Speed and Visibility	30
2.7	Generating Inputs		32
	2.7.1	Dealing with Configurations and Settings	32
	2.7.2	Generating and Driving Digital Control	33
2.8	Checking Analog Functionality		38
	2.8.1	Comparing Two Values	38
	2.8.2	Triggering a Check on Control Changes	40
	2.8.3	Measuring Signal Timing	43
	2.8.4	Comparing a Value to a Threshold	45
	2.8.5	Checking Frequency Response	47
2.9	Using Assertions		49
	2.9.1	Checking Input Conditions	49
	2.9.2	Verifying Local Invariants	50
	2.9.3	Limitations on Assertion Checking of Analog Properties	50
	2.9.4	Dealing with Different Modeling Styles	50
2.10	Clocks, Resets and Power Controls		51
	2.10.1	Driving Clocks	51
	2.10.2	Resets	51
	2.10.3	Power-Up and Power-Down Sequences	51
2.11	Analog Model Creation and Validation		53
2.12	Integrating the Test Environment		54
	2.12.1	Connecting the Testbench	54
	2.12.2	Connecting to Electrical Nodes	55
	2.12.3	System-Level Parameters and Timing	56
	2.12.4	Supporting Several Model Styles In A Single Testbench	57
	2.12.5	Interfacing Between Real and Electrical Signals	59
	2.12.6	Creating Run Scripts and Other Support Files	61
	2.12.7	Recommended Directory Structure	62
2.13	Closing the Loop Between Regressions and Plan		63
	2.13.1	Implementation of Coverage for Analog	63
	2.13.2	Updating the Verification Plan With Implementation Data	64
2.14	Regression Runs for Analog IP		65
	2.14.1	Single Simulation Runs	65
	2.14.2	Regressions—Running Multiple Test Cases	67
2.15	Moving Up to the SoC Level		69
	2.15.1	Mix-and-Match SoC-Level Simulation	69
	2.15.2	Updating the SoC-Level Test Plan	70

		2.15.3	Integrating Into the SoC-Level Testbench 70
	2.16	UVM-MS Universal Verification Blocks 72	
		2.16.1	Wire Verification Component 72
		2.16.2	Simple register UVC 78
		2.16.3	Analog to Digital Converter (ADC) UVC 82
		2.16.4	Digital to Analog Converter (DAC) UVC 84
		2.16.5	Level Crossing Monitor 86
		2.16.6	Ramp Generator and Monitor 89
	2.17	Summary 94	

3 Low-Power Verification with the UVM 97

	3.1	Introduction 97	
		3.1.1	What is Unique about Low-Power Verification 98
		3.1.2	Understanding the Scope of Low-Power Verification 99
		3.1.3	Low-Power Verification Methodology 100
		3.1.4	Understanding Low-Power Verification 102
	3.2	Understanding Low-Power Design and Verification Challenges 102	
		3.2.1	How Low-Power Implementations Are Designed Today 102
		3.2.2	Challenges for Low-Power Verification 103
		3.2.3	Low-Power Optimization 104
		3.2.4	Low-Power Architectures 105
		3.2.5	Low-Power Resources 111
	3.3	Low-Power Verification Methodology 111	
	3.4	Low-Power Discovery and Verification Planning 113	
		3.4.1	Low-Power Discovery 113
		3.4.2	Verification Planning 113
		3.4.3	System-Level Planning 113
		3.4.4	Hierarchical Planning 117
		3.4.5	Domain-Level Verification Planning 118
		3.4.6	A Note on Verifying Low-Power Structures 119
		3.4.7	Recommendations for Designs with a Large Number of Power Modes 119
	3.5	Creating a Power-Aware UVM Environment 121	
		3.5.1	Tasks for a Low-Power Verification Environment 121
		3.5.2	Solution: Low-Power UVC 122
		3.5.3	UVC Monitor 124
		3.5.4	LP Sequence Driver 126
		3.5.5	UVM-Based Power-Aware Verification 128
	3.6	Executing the Low-Power Verification Environment 128	
		3.6.1	LP Design and Equivalency Checking 128
		3.6.2	Low-Power Structural and Functional Checks 129

		3.6.3	Requirements for Selecting a Simulator and Emulator 130
		3.6.4	Advanced Debug and Visualizations 132
		3.6.5	Automated Assertions and Coverage 135
		3.6.6	Legal Power Modes and Transitions 135
		3.6.7	Automatic Checking of Power Control Sequences 135
		3.6.8	Verification Plan Generated from Power Intent 136
	3.7	Common Low-Power Issues 138	
		3.7.1	Power-Control Issues 138
		3.7.2	Domain Interfaces 139
		3.7.3	System-Level Control 140
	3.8	Summary 141	

4 Multi-Language UVM 143

	4.1	Overview of UVM Multi-Language 143	
	4.2	UVC Requirements 146	
		4.2.1	Providing an Appropriate Configuration 146
		4.2.2	Exporting Collected Information to Higher Levels 146
		4.2.3	Providing Support for Driving Sequences from Other Languages 146
		4.2.4	Providing the Foundation for Debugging of All Components 146
		4.2.5	Optional Interfaces and Capabilities 147
	4.3	Fundamentals of Connecting *e* and SystemVerilog 147	
		4.3.1	Type Conversion 147
		4.3.2	Function Calls Across Languages 150
		4.3.3	Passing Events Across Languages 153
	4.4	Configuring Messaging 154	
	4.5	*e* Over Class-Based SystemVerilog 154	
		4.5.1	Environment Architecture 155
		4.5.2	Configuration 156
		4.5.3	Generating and Injecting Stimuli 160
		4.5.4	Monitoring and Checking 168
	4.6	SystemVerilog Class-Based over *e* 171	
		4.6.1	Simulation Flow in Mixed *e* and SystemVerilog Environments 173
		4.6.2	Contacting Cadence for Further Information 173
	4.7	UVM SystemC Methodology in Multi-Language Environments 174	
		4.7.1	Introduction to UVM SystemC 174
		4.7.2	Using the Library Features for Modeling and Verification 175
		4.7.3	Connecting between Languages using TLM Ports 182
		4.7.4	Example of SC Reference Model used in SV Verification Environment 188
		4.7.5	Reusing SystemC Verification Components 191
	4.8	Summary 193	

5 Developing Acceleratable Universal Verification Components (UVCs) ... 195

- 5.1 Introduction to UVM Acceleration ... 195
- 5.2 UVC Architecture ... 196
  - 5.2.1 Standard UVC Architecture ... 196
  - 5.2.2 Active Agent ... 196
  - 5.2.3 Passive Agent ... 197
  - 5.2.4 Acceleratable UVCs ... 197
- 5.3 UVM Acceleration Package Interfaces ... 201
  - 5.3.1 uvm_accel_pipe_proxy_base Task and Function Definitions (SystemVerilog) ... 202
  - 5.3.2 uvm_accel_pipe_proxy_base Task and Function Definitions (*e*) ... 204
- 5.4 SCE-MI Hardware Interface ... 205
  - 5.4.1 SCE-MI Input Pipe Interface ... 206
  - 5.4.2 SCE-MI Output Pipe Interface ... 206
- 5.5 Building Acceleratable UVCs in SystemVerilog ... 207
  - 5.5.1 Data Items ... 207
  - 5.5.2 Acceleratable Driver (SystemVerilog) ... 209
- 5.6 Building Acceleratable UVCs in *e* ... 215
  - 5.6.1 Data Items ... 215
  - 5.6.2 Acceleratable Driver (*e*) ... 216
- 5.7 Collector and Monitor ... 219
- 5.8 Summary ... 219

6 Summary ... 221

The Authors ... 223

Index ... 227

List of Figures

Figure 1-1	Verification Plan for the UART Block	3
Figure 1-2	Executable vPlan with Coverage	4
Figure 1-3	vPlan with Metrics	5
Figure 1-4	Details of a Kit Testbench Architecture	6
Figure 1-5	Cluster-Level Testbench for the Kit APB Subsystem	6
Figure 1-6	MDV Changes an Open-Loop Verification Process into a Closed-Loop Process	7
Figure 1-7	Architecture Diagram of the Kit Platform Including a Power Control Module	9
Figure 1-8	Architectural Overview of a Chip-Level Testbench and its Corresponding Chip-Level vPlan	10
Figure 1-9	Analysis of Coverage Contribution for Each Run of a Regression Against a Specific vPlan Feature	11
Figure 1-10	Example of MDV Applied to a Real Customer Project	12
Figure 1-11	The Continuum that Exists between Hardware/Software, Verification, and Validation	13
Figure 2-1	Applying UVM-MS at the IP and SoC Level	17
Figure 2-2	Verification Flow at the IP Level	18
Figure 2-3	Verification at the SoC Level Reusing IP Verification Elements	19
Figure 2-4	Enterprise Planner Used to Capture Analog Property	21
Figure 2-5	Plan Implementation Notes	22
Figure 2-6	A Noisy Sine Wave Source Created by Composing Two Signal Sources	24
Figure 2-7	Threshold Checking Example	25
Figure 2-8	Architecture of a UVM-MS Environment	26
Figure 2-9	Architecture of a Power Controller Driven by Sequence	34
Figure 2-10	An Example of a Top-Level Sequence Controlling Three UVCs	37
Figure 2-11	A Checker Comparing DUT's Input and Output to Compute Gain	39
Figure 2-12	Timing Sequence for the Gain Checker	39
Figure 2-13	Triggering Checks Automatically Upon Control Value Change	41
Figure 2-14	Checking for Signal Timing	43
Figure 2-15	Timing a Measurement of a Dynamic Signal for Checking DC Bias at Time t-max	43
Figure 2-16	Threshold Crossing Checker	45
Figure 2-17	A Simple Power-Up Spec	52
Figure 2-18	The Effects of Sampling a Quantized Signal	56
Figure 2-19	Connect Module Resolution Derived from Supply Voltage	60
Figure 2-20	Unmapped Analog Coverage Data from Regression Runs	64
Figure 2-21	Initial vPlan for Analog Components	65
Figure 2-22	Running Regression using Incisive Enterprise Manager	67
Figure 2-23	Load vPlan for Analog IP	68

List of Figures

Figure 2-24	Analog Coverage	68
Figure 2-25	Properties Controlled and Measured by the Wire UVC	72
Figure 2-26	Structural Diagram of the Wire UVC	73
Figure 2-27	Structure Diagram of a Generated Register UVC	78
Figure 2-28	Voltage and Time Points Controlled and Measured by the Ramp UVC	89
Figure 2-29	Structural Diagram of the Ramp UVC	90
Figure 2-30	Ramp Waveform Resulting Generated by Above Sequence	94
Figure 3-1	Low-Power Verification Flow Diagram	100
Figure 3-2	Chart of Low-Power Complexity Due to Power Modes	103
Figure 3-3	Example Low-Power Design	105
Figure 3-4	Example Multiple Supply voltages Domain	107
Figure 3-5	Example Power-Shutoff Domain	109
Figure 3-6	Low-Power Verification Methodology	112
Figure 3-7	System-Level Power Plan	115
Figure 3-8	Power-Mode Selection of the Verification Plan	116
Figure 3-9	Low-Power UVM Environment	123
Figure 3-10	Overview of Low-Power Visualizations	133
Figure 3-11	Waveform of Power-Shutoff Corruption	134
Figure 3-12	Waveform of Isolation	134
Figure 3-13	Tracing a Signal Includes Low-Power Signals	135
Figure 3-14	Power-Up/Down Sequence	136
Figure 3-15	Automated Verification Plan	137
Figure 3-16	Oscillations on Control Signal for PSO	138
Figure 3-17	Bus Hang Due To Incorrect Isolation	140
Figure 4-1	UVC Architecture	145
Figure 4-2	Sample Environment: UART Module UVC	155
Figure 4-3	Typical Module/System UVC Architecture	156
Figure 4-4	Layering an *e* Proxy Sequencer on Top of the SV-UVC Sequencer	161
Figure 4-5	UVC — SystemVerilog API over *e*	172
Figure 4-6	Configurable DUT Instantiated under the Testbench	178
Figure 4-7	SC_MODULE DUT Separate from the Testbench	178
Figure 4-8	Multi-Language Port Connection	182
Figure 4-9	Verification with Reference Model	188
Figure 4-10	Reusing a SystemC Driver	191
Figure 5-1	Standard UVC Architecture	196
Figure 5-2	Components of the HVL and HDL Partitions	198
Figure 5-3	Acceleratable UVC Architecture	199
Figure 5-4	Acceleratable Transactors	200
Figure 5-5	Transactor Example with Input and Output Channels	200
Figure 5-6	Packed Implementation of Data Item yamp_transfer	208
Figure 5-7	Packed Implementation of Data Item yamp_transfer	216

List of Tables

Table 2-1	Recommended Structure for Top-Level Project Directory	62
Table 2-2	Configuration Parameters for the Wire UVC	75
Table 2-3	Sequence Items Provided for the Wire UVC	76
Table 2-4	dms_register UVC Sequence Item Parameters	80
Table 2-5	Configuration Parameters for the Threshold Monitor UVC	88
Table 2-6	Available Port Connections for Integrating the Threshold Monitor	88
Table 2-7	Configuration Parameters for the Ramp UVC	92
Table 2-8	Sequence Items Provided for Ramp UVC	93
Table 3-1	Most Common Techniques for Low-Power Optimization	104
Table 4-1	UVC Layers	144
Table 4-2	Recommended SystemVerilog Adapter Configuration	148
Table 5-1	SystemVerilog uvm_accel_input_pipe_proxy Task and Function Definitions	203
Table 5-2	SystemVerilog uvm_accel_output_pipe_proxy Task and Function Definitions	204
Table 5-3	*e* uvm_accel_input_pipe_proxy Unit Definition	204
Table 5-4	*e* uvm_accel_input_pipe_proxy Task and Function Definitions	205
Table 5-5	*e* uvm_accel_output_pipe_proxy Task and Function Definitions	205

List of Code Examples

Example 2–1 *e* Code Implementing a Directly Measured Real Coverage Item 27
Example 2–2 *e* Code Implementing Delayed Coverage Collection 28
Example 2–3 *e* Code Collecting Coverage Properties of a Dynamically Changing Signal 29
Example 2–4 *e* Code Using when Construct to Declare Different Testbench Configurations 31
Example 2–5 *e* Code for Generating and Applying Configuration 32
Example 2–6 *e* Code Example—Power Control Interface 34
Example 2–7 *e* Test File Defining a Sequence that Sets Up Two Signal Generators and Measures a Resulting Signal 36
Example 2–8 *e* Code Implementing a Top-Level Sequence Skeleton 37
Example 2–9 *e* Code Example of Gain Checker 39
Example 2–10 *e* Code Activating Amplitude Measurement upon Change in Control Registers 41
Example 2–11 This Checker Doesn't Assume Input and Output Samples are Synchronized 42
Example 2–12 *e* Code for Two Simple Checkers Verifying Voltage/Timing 44
Example 2–13 *e* Code Using Threshold Crossing Monitors to Check Voltage-Level Range, Ignoring an Initial Power-Up Period 46
Example 2–14 *e* Code Featuring Top-Level Sequence that Applies a Band of Frequencies Around a Central Operating Frequency 47
Example 2–15 *e* Code Example Implementing a Predictor 48
Example 2–16 PSL Formula for Checking Voltage Levels at Specific Point in Time 49
Example 2–17 PSL Formula for Conditional Checking at Specific Point in Time 49
Example 2–18 PSL Formula that Checks for an Invariant Condition 50
Example 2–19 PSL Verification Unit that is Bound to diFferent Modules Depending on the Model Used 50
Example 2–20 *e* Code Example Combining Checking and Coverage of a Simple Power Sequence 52
Example 2–21 *e* Code Example, Connecting to Electrical Nodes 55
Example 2–22 *e* Code Example of Connections to Real Register and Real Wire (wreal) 55
Example 2–23 *e* Code Example of a Port Connected to an Expression in Verilog-AMS Domain 56
Example 2–24 Verilog Code—Netlist Elements and Real-Number Modeling 57
Example 2–25 *e* Code Example—Dynamic Sub-Type of Testbench 58
Example 2–26 Command-Line Options for irun 59
Example 2–27 Using the Simulation Control File amscf_spice.scs to Control Connect Module Features .60
Example 2–28 A Simple run_sim File Invoking irun 61
Example 2–29 Minimal run.tcl File Collecting Waveforms During Run 61
Example 2–30 A Sample run.f File Used to Define the Build and Run Process 61
Example 2–31 *e* Code for Cover Groups Implemented in Testbench 63
Example 2–32 Commands for Invoking a Simulation Run 65
Example 2–33 A Simulation Run File Example 66
Example 2–34 An Example Build and Run Command File 66

List of Code Examples

Example 2-35 Verilog Instantiation of Wire Modules, Both Single Wire and Differential Pair Interfaces—Step 1 in the Integration Process 74
Example 2-36 *e* Code Example for Instantiating, Constraining and Connecting a Wire Unit—Steps 2-5 in the Integration Process 74
Example 2-37 *e* Code Example of Integrating a Wire Driver Into A Top-Level Sequence—Step 6 in the Integration Process 74
Example 2-38 *e* Code Example of A Top-Level Sequence Driving A Wire UVC Source and Monitor. . . . 76
Example 2-39 *e* Code Showing the Implementation of an Input Method Port for Reading Measured Wire Data 77
Example 2-40 *e* Code of the Default Cover Group Collected By the Wire Monitor 77
Example 2-41 Syntax of Interface Definition Used By the dms_register UVC 79
Example 2-42 *e* Code Example Defining an Interface With Two Registers—Step 1 in the Integration Process 79
Example 2-43 *e* Code Example Instantiating a Control Register Interface and Connecting It—Step 2 in the Integration Process 80
Example 2-44 *e* Code Example Connecting the Control Interface Driver to the Top-Level Sequence Driver—Step 3 in the Integration Process 80
Example 2-45 *e* Code Example Of Top-Level Sequence Configuring Control Registers Using Sequence Items 81
Example 2-46 *e* Code Back-Door API For Accessing Register Values 81
Example 2-47 *e* Code Example Using Back-Door Access Method To Register Values 81
Example 2-48 Instantiating the ADC Block Using Simple Port Connections 83
Example 2-49 *e* Code Example Using ADC Block Output In A Monitor 83
Example 2-50 Instantiating the DAC Block, Using Simple Port Connections 85
Example 2-51 *e* Code Example Using DAC Block Output In A Monitor 85
Example 2-52 Verilog Instantiation Of the Threshold Monitor—Step 1 in the Integration Process 87
Example 2-53 *e* Code Example For Instantiating and Constraining a Threshold Unit—Steps 2 and 3 in the Integration Process 87
Example 2-54 *e* Code Hooking Up the Monitor Event Port Output—Step 4 in the Integration Process . . 87
Example 2-55 Verilog Instantiation Of Ramp Modules, One For Driving and One For Monitoring the Output—Step 1 in the Integration Process 90
Example 2-56 *e* Code Example for Instantiating, Constraining and Connecting the Ramp UVCs—Steps 2 and 3 in the Integration Process 91
Example 2-57 *e* Code Hooking Up the Monitor Output—Step 4 in the Integration Process 91
Example 2-58 *e* Code Example Of Integrating a Ramp Driver Into a Top-Level Sequence—Step 5 in the Integration Process 92
Example 2-59 *e* Code Example of a Top-Level Sequence Driving a Wire UVC Source and Monitor 93
Example 4-1   Calling a SystemVerilog function from *e* 150
Example 4-1   Calling a Method from SystemVerilog 151
Example 4-1   *e* Code Sensitive to SystemVerilog Events 153

Preface

The statement "it's a digital world" is a gross simplification of reality. There is no doubt that digital content in electronics design is growing geometrically with Moore's law. However, the addition of integrated analog functions, third-party IP, power management, and software is creating an exponentially-scaled verification problem. Furthermore, this scaling virtually guarantees that inconsistent verification approaches will introduce problems because every project involves multiple, often third-party, global project teams. The convergence of new functionality, exponentially-scaled verification, and distributed teams at advanced nodes such as 20 nm creates operational pressure as companies try to profitably meet the verification requirements of modern Systems on Chip (SoCs). Following the conventional engineering approach to address a huge system, industry-leading teams established a controlled, working base for their verification environment and then built upon that base. Those teams built and selected the Universal Verification Methodology (UVM) standard from Accellera as that base because it is architected to scale with the growing size of digital verification. As the UVM base expands, verification managers and team leaders are now asking "what's next?"

The answer to that question depends on the requirements that SoC verification adds on top of digital verification. An SoC that a team assembles from internal and 3rd-party hardware IP may be primarily digital, but it clearly needs to integrate verification IP (VIP) from multiple sources. If the commercial differentiation of the SoC depends on unique features, then mixed signal and low power are likely to be critical next steps. Verification capacity and the means to automate a comprehensive verification plan are the areas of focus when the SoC is so large and complex that it can only be implemented in the latest available node.

Of these requirements, VIP integration is the most common. Today, the UVM language standard is IEEE 1800 SystemVerilog, but many UVM users need to access VIP written in IEEE 1647 *e* or IEEE 1666 SystemC. Because many SystemVerilog verification teams would like to easily integrate VIP or models written in different languages, it is essential that the UVM is extended to support multi-language, interoperable verification environments. Doing so builds on the reuse inherent in the UVM and preserves the quality coded into the existing VIP.

Almost as pervasive as VIP for SoC integration are low-power and mixed-signal needs. Starting as a trend in the deep-submicron era and continuing through 20 nm and below, the drive toward further integration assures that nearly every SoC will have this extra complexity. For low-power design, the change becomes evident in the move from simple clock gating to power shutoff, voltage scaling, and other methods that frequently have multiple hardware and software controls. As a result, the handful of directed tests that once verified the power modes are no longer sufficient to assure proper power response in tens or hundreds of thousands of sequences that verify the function of the SoC. A better approach is to make the base UVM environment power aware, thereby fusing functional and power verification. Similarly, the hardware function of the SoC itself is increasingly dependent on both analog and digital circuits, so mixed-signal verification needs to be added to this fused solution. Adding these extra capabilities to the verification task raises analog-specific questions about simulation speed, modeling abstraction, coverage, and assertions. While some standards in the mixed-signal space, including IEEE 1800 SV-DC (analog modeling for SystemVerilog), remain in-flight at the time this book was written, many of the capabilities to implement a working methodology do exist because project teams are fusing low-power, mixed-signal, and UVM functional verification for 20 nm designs today.

Scaling is third requirement. It is pervasive at 20 nm, but becomes evident in the increased complexity nearly every team deals with in each project. Verification engineers assembling the full SoC for verification are seeing turn-around time for big software-based verification runs slowing significantly as their projects grow. While hardware-based verification does not suffer the same effect, it appears to be out-of-reach technically for many teams. By creating a connection to the UVM for the testbench, hardware-based acceleration becomes a more attractive option for software-based verification teams. Of course, those teams can only reach the point of integrated SoC acceleration by standing on the shoulders of the block and subsystem verification work already completed. Teams know they can stand firmly on this quality base if they have followed a metric-driven verification methodology (MDV) that captures metrics at every stage and maps these metrics back to their verification plan.

Consumers may perceive that "it's a digital world," but these advanced verification topics speak to the magic that goes on under the hood of every SoC. As verification engineering managers and team leaders, we know that MDV, multi-language VIP, low-power, mixed-signal, and acceleration topics are converging at 20 nm and beyond; but we don't want to create whole new methodologies for each one. The authors of this book realized this, and selected the Accellera UVM standard as the common base from which to offer solutions that leverage reuse and raise team-level productivity. That's why we have written this book—not only for verification engineers familiar with the UVM and the benefits it brings to digital verification, but also for verification engineers who need to tackle these advanced tasks. Though the solutions in this book are not standardized, most of them are available through open-source code. For all of you, the material in this *Advanced Verification Topics* book is provided as a means to stay productive and profitable in the face of growing verification complexity.

—Adam Sherer, Cadence Product Marketing Director

How to Use This Book

Advanced Verification Topics is divided into several sections; each focusing on a different specialization. It is recommended that the reader be familiar with the UVM. The best resource for that is *A Practical Guide to Adopting the UVM*, by Sharon Rosenberg and Kathleen Meade.

The following is an introduction to each section of *Advanced Verification Topics*.

- **Introduction to Metric-Driven Verification** integrates all of the verification specializations. Working in a cycle of continuous refinement, the verification team builds and maintains a comprehensive verification plan then gathers metrics from each specialization. As the plan converges, the verification team has a clear, continuous understanding of where the project is on the path to silicon realization.

- **UVM and Metric-Driven Verification of Mixed-Signal Designs** brings digital verification techniques to the analog world. Though this preface begins with "The statement it's a 'digital world' is a gross simplification," every engineer knows that it is actually a digital island in an analog world. As that island has grown, directly testing each analog interface and the interaction among them is no longer sufficient. This section explains how digital verification techniques can be applied to, and integrated with, analog design to extend the UVM for mixed-signal applications.

- **Low-Power Verification with the UVM** elevates low-power from a niche-directed check. When the design has two power modes, it's reasonable for a separate team to write a few tests and declare the low-power aspects verified. When the design has multiple power domains with multi-level voltages and both hardware and software triggers, there may be tens of thousands of possible power modes and only a handful of legal modes—virtually guaranteeing bug escapes if a directed test methodology is employed. This section describes a comprehensive low-power verification methodology integrated with the mainstream digital verification task, and executed with both the UVM and the verification planning processes defined by Metric-Driven Verification.

- **Multi-Language UVM** provides a solution to the business/technical metric that enables alliances to form and reform while maintaining engineering efficiency. It defines parallel implementations of the UVM in SystemVerilog, *e*, and SystemC enabling each alliance contributor to make language decisions best suited to their subsystem needs, yet seamlessly integrated to verify the complete SoC. All of this is done without recoding any of the verification IP. That is the definition of efficient reuse and is only possible because of the multi-language architecture of the UVM.

- **Developing Acceleratable Universal Verification Components (UVCs)** defines how to best verify those huge digital islands. When simulations run for days in order to process enough clock cycles to verify system functions, hardware acceleration is needed. As with the other specializations, this one steps up from a necessary technology to an efficient methodology for multi-supplier alliances, and enables verification convergence and silicon realization when it:
    - Reuses the stimulus defined in the UVM simulation environment
    - Communicates using the Accellera SCE-MI standard
    - Supplies data for Metric-Driven Verification

1 Introduction to Metric-Driven Verification

This chapter introduces the Cadence® Incisive® Verification Kit as a golden example of how to maximize verification effectiveness by applying metric-driven verification (MDV) in conjunction with the Universal Verification Methodology (UVM). MDV provides an overarching approach to the verification problem by transforming an open-ended, open-loop verification process into a manageable, repeatable, deterministic, and scalable closed-loop process. Through this transformation, verification project teams and managers greatly increase their chances of consistently delivering working designs at improved quality levels in less time with fewer human resources.

1.1 Introduction

MDV is a key to Silicon Realization, part of the greater EDA360 vision. EDA360 helps both design creators and design integrators close the "productivity gap" (through improved approaches to design, verification, and implementation) as well as the "profitability gap" (by providing new capabilities for IP creation/selection/integration and system optimization). Silicon Realization represents everything it takes to get a design into working silicon, including the creation and integration of large digital, analog, and mixed-signal IP blocks. Advanced verification capabilities, such as MDV, are a must.

The functional verification landscape has changed beyond all recognition over the last 10 years, and while design paradigms have become mature and stable, verification methodologies and technologies have continued to evolve and new flows and tools are still being invented. Against this rapidly changing background, designs have been steadily growing with more IP and more complex IP being integrated into larger and larger SoCs.

Realizing silicon in the face of these challenges requires new approaches and a very flexible workforce capable of adapting and changing on a regular basis. This paper outlines a new approach to managing verification complexity—metric-driven verification—and a valuable resource to back up this new approach, the Cadence Incisive Verification Kit, which together will enable you to plan and embrace new methodologies and approaches in a safe and controlled way. The kit provides clear and realistic examples at each stage to guide you and your teams in their new projects.

Silicon Realization is achieved when metrics can be applied to design intent, when **users** can work more productively because they are working at a higher level of abstraction, and when successive refinements converge to achieve verification closure. MDV manages intent, abstraction, and convergence—leading to greater predictability, productivity, and eventually profitability.

1.2 Failing to Plan = Planning to Fail

Any management process needs clear and measurable goals, and verification is no exception. Failing to capture these goals at the outset of a project means that there is no clear definition against which to measure either progress or closure. You can only gauge improvement in what you can clearly measure. Historically this was a problem with directed testing. Huge lists would be drawn up to define the verification process, but these lists were not executable or maintainable. This open-ended nature led to big project slips and huge stresses on project teams.

In addition, there is often confusion about the definition of what constitutes a verification plan and what constitutes a test plan. Let's make our definition clear right away: a test plan defines a fully elaborated list of tests that will be created and the completion of this list defines completion of verification. A test plan alone, however, tends to be a poor mechanism because it already contains decisions about the execution of the process. In contrast, a verification plan captures the "what" that needs to be verified but does not define the execution—the "how." This is a crucially important distinction as it is expected that verification may be done using multiple technologies by multiple people.

We refer to the "what" as the "features" or "goals" that need to be verified—it is the design intent. When planning a project, you should not presume the underlying tool by which a feature is verified. Verification will most likely come from several sources, and the results are represented by various forms of coverage and other metrics.

A verification plan should capture the goals of the verification process such that the sign-off criteria are clearly identified as well as milestones along the way. An example of a milestone would be register-transfer level (RTL) code freeze, which a team might define as having 75% of all features covered by that time.

In addition to capturing goals up-front, a verification plan should be executable. In other words, it is essential that progress toward project closure can be easily and automatically measured. Typically, the goals get refined as a project progresses, which means the verification plan is a living document that matures over the lifetime of a project.

Figure 1-1 on page 3 shows a sample of a verification plan for the UART block in the Cadence Incisive Verification Kit environment.

Figure 1-1 Verification Plan for the UART Block

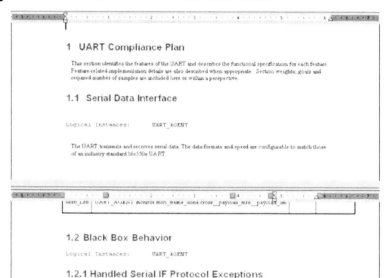

As the sample shows, the verification plan can be a regular Microsoft Word document that uses template heading styles to identify structure and specific features. This is easy to create in Microsoft Word or using Incisive Enterprise Planner as the front-end editor of the plan. The verification plan also identifies abstract features and hierarchies of features that may closely resemble the hierarchy of the specification. While not mandatory, a verification plan is often a useful convenience that helps designers and verification engineers communicate. Verification plans can also include other (sometimes commercially supplied) verification plans. In the case of the Incisive Verification Kit, the UART has an AMBA® APB interface and, therefore, the overall kit plan includes an APB verification plan. This mechanism enables reuse of plans that relate to standard interfaces or perhaps blocks of IP that are reused across multiple projects.

Figure 1-2 on page 4 shows an executable verification plan (vPlan) with coverage mapped onto it. This approach allows resource decisions to be made in an informed way such that progress toward closure proceeds according to the original plan. The coverage is accumulated from any number of runs of a verification flow and from a variety of verification engines (formal, simulation, acceleration, or emulation).

Figure 1-2 Executable vPlan with Coverage

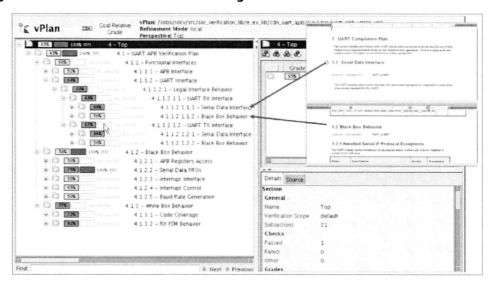

1.3 Metric-Driven Verification

Over the past few years, the term "coverage-driven verification" has become widely adopted to refer to the gathering of simulation coverage in all of its various forms, traditionally code and functional coverage. This approach has many limitations that affect its usability and scalability. One major limitation is that it does not include checking or time-based aspects, which are essential in defining the closure criteria.

Metric-driven verification (MDV) broadens the scope of what is captured and measured to include checks, assertions, software and time-based data points that are encompassed in the term "metrics."

The second enhancement MDV offers over coverage-driven verification is the ability to create feature hierarchies by using an executable verification plan (vPlan). This helps manage the wealth of data captured by all the tools involved in the execution of a verification project.

The third enhancement is support for parallel verification execution to optimize verification throughput. One of the primary objectives of MDV is to reduce the dependency on the one resource that is not scalable—the verification engineer. By enabling use of parallel processing via machines through automation, MDV improves productivity and reduces the number of non-productive tests. Figure 1-3 on page 5 contains a vPlan comprising metrics from a number of different environments and different verification engines.

Figure 1-3   vPlan with Metrics

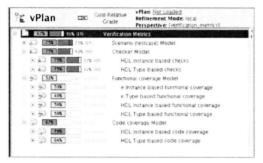

We can see how metrics are gathered at the UART level from multiple runs of both formal (static) and simulation (dynamic) verification.

1.4   Building Strong Testbench Foundations

All of the high-level capabilities of MDV that deliver the abstracted and filtered view of the verification process would amount to nothing without a common framework under which verification environments could be constructed and reused. The Universal Verification Methodology (UVM) is the Accellera standard for the construction of verification environments. The UVM is based on the Open Verification Methodology (OVM) 2.1.1 release (the OVM itself an evolution from the *e* Reuse Methodology (*e*RM), which has been in widespread use since 2002 on thousands of projects). The UVM supports all of the important verification and modeling languages customers demand and enjoys industry-wide support.

Given the investment needed in training and development of advanced verification environments, it is reassuring to know that by adopting the UVM you are part of a global community working toward common verification goals. MDV builds on top of the UVM, enables the management of complex SoC developments, and focuses on the key challenges rather than issues of verification environment implementation or reuse.

The Incisive Verification Kit completely adopts the UVM and includes real-world testbench examples in both *e* and SystemVerilog. At the block level, the kit shows how MDV can be applied to the verification of a UART design. It shows how a verification plan can be constructed up-front, how a UVM environment can be rapidly constructed using Universal Verification Components (UVCs), and how simulations can be run and coverage annotated against the verification plan to monitor and manage the verification process.

Figure 1-4 on page 6 details the architecture of one of the kit testbenches, showing the hook-up of the UVCs to the UART design under test (DUT).

Introduction to Metric-Driven Verification

Figure 1-4 Details of a Kit Testbench Architecture

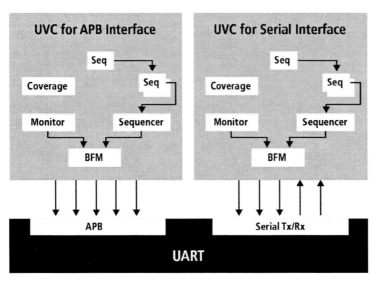

One of the major strengths of the UVM is its approach to reuse. Improving productivity is all about writing complex testbench elements—once and only once— and thereafter reusing these components as widely as possible. One of the dimensions of reuse engineered into the UVM is from block to cluster.

Figure 1-5 shows the cluster-level testbench for the kit APB subsystem, clearly illustrating the large amount of reused content. Notice the reused serial interface UVCs and the reused APB UVCs.

Figure 1-5 Cluster-Level Testbench for the Kit APB Subsystem

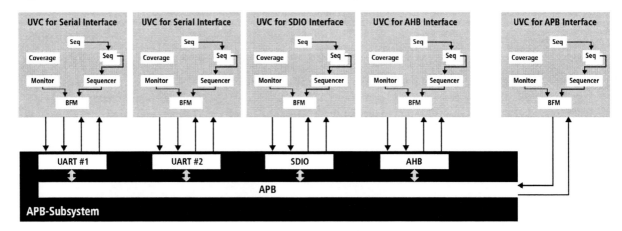

One of the key purposes of the kit is to accelerate understanding of UVM concepts, and testbench architecture diagrams like this are available as clickable demos in the kit. These diagrams enable you to quickly familiarize yourself with both the methodology and language usage in the context of a realistic example.

This modular, layered approach also forms the basis for reuse. In addition to the creation of plug-and-play hardware and software verification components that can be reused from block to cluster to chip to system, UVM components can also be reused across multiple projects and platforms.

As there is a frequent need for verification components for standard interfaces such as USB, PCI Express, AMBA AXI, and so on, Cadence has created a comprehensive portfolio of UVM-compliant commercial verification IP (VIP) components. These components are of particular interest when it comes to "compliance testing" (ensuring that the design's interfaces fully comply with their associated specifications). Furthermore, each of these VIP components comes equipped with its own pre-defined verification plan that can be quickly and easily incorporated into a master plan. (A subset of this VIP portfolio is demonstrated within the Incisive Verification Kit.)

1.5 Simulation Isn't the Only Way

MDV uses a generic approach that doesn't mandate any specific verification execution engine; in other words simulation isn't the only tool you may choose to use for establishing design correctness. Figure 1-6 illustrates how MDV changes an open-loop verification process into a closed-loop process through a Plan-Construct-Execute-Measure cycle using the results of multiple verification engines.

Figure 1-6   MDV Changes an Open-Loop Verification Process into a Closed-Loop Process

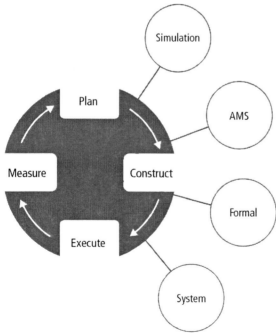

For example, formal tools are another means by which design correctness can be established. Formal properties in the form of assertions are captured, typically written manually by the user, although automatically extracted assertions and assertion libraries can also accelerate the verification process. Formal tools attempt to exhaustively prove that, for all possible circuit states, the assertion is true, or to demonstrate how the assertion can fail. The Incisive Verification Kit provides examples of these flows and shows how the assertion checks and coverage points are easily included in the verification plan. The results from the formal process are then annotated onto the plan.

MDV seamlessly provides verification engineers and managers with a consistent view of the verification progress, regardless of where the results are coming from. While exhaustively proving assertions is a goal, there are frequently occasions where the tools need additional guidance to complete their proof. Typically, the user writes additional assertions that define constraints. These narrow the scope of the formal proof, and the tools are then able to complete the proof. The question is: how do you know that the assumptions in these constraints are valid?

Assertions are not only used for formal proof, but will also execute in simulation. By reusing the constraints in your simulation runs, you have a means of cross-checking whether the assumptions you made still hold true in your simulation models. This isn't a watertight mechanism, as users may not exhaustively simulate the environment in all possible scenarios. However, if you adopt constrained-random simulation with the UVM, you increase your chances of hitting the corner-case where assumptions may have been incorrect.

MDV provides a mechanism for users to include the coverage of the constraints, which, at a minimum, will identify that a constrained condition was triggered. You always have visibility into failures, so the coverage you get gives you a useful cross-check of these constraints. These techniques from both the planning and execution side are included as workshops within the kit.

The Cadence MDV flow supports the use of formal tools to prove several types of coverage unreachable. This is very valuable in the late stages of your project, when a small percentage of coverage remains and is proving hard to hit. It may be that some of the remaining coverage is simply impossible to reach no matter how good your UVM testbench or how long you let it run. Any coverage proved unreachable by formal can annotated into the vPlan to drive verification convergence.

1.6   Low Power isn't Just the Designer's Problem

Low-power design is no longer a "nice to have," but mandatory even for non-mobile applications, and a crucial "must have" for all "green" products. But as a verification engineer, why should you care about low power?

In the past, the verification engineer would verify that the RTL implementation precisely matched the design specification, and while that still holds true, it is only part of the problem. With the addition of new low-power techniques that capture power intent in side files, the "golden" design is no longer just the RTL, but is a combination of the RTL plus the power intent. While the power intent defines the physical implementation details, there is also a power control module (PCM) that is—almost without exception—under software control.

The low-power verification challenge therefore has two new dimensions. Firstly, the physical implementation changes can introduce functional bugs as a side effect. These may only manifest themselves during corner-case scenarios. Secondly, the interaction between the power management software and the hardware requires careful verification to ensure that all power modes and power-mode transitions have been verified in all operating circumstances. Even a design with just a few pieces of IP that can be powered down can have thousands of operational states and power transitions that must be exercised.

The Incisive Verification Kit provides comprehensive material and examples covering verification of a hardware platform containing a realistic set of power management features.

Figure 1-7 shows a simplified architecture diagram of the kit platform including a power control module (PCM).

Figure 1-7 Architecture Diagram of the Kit Platform Including a Power Control Module

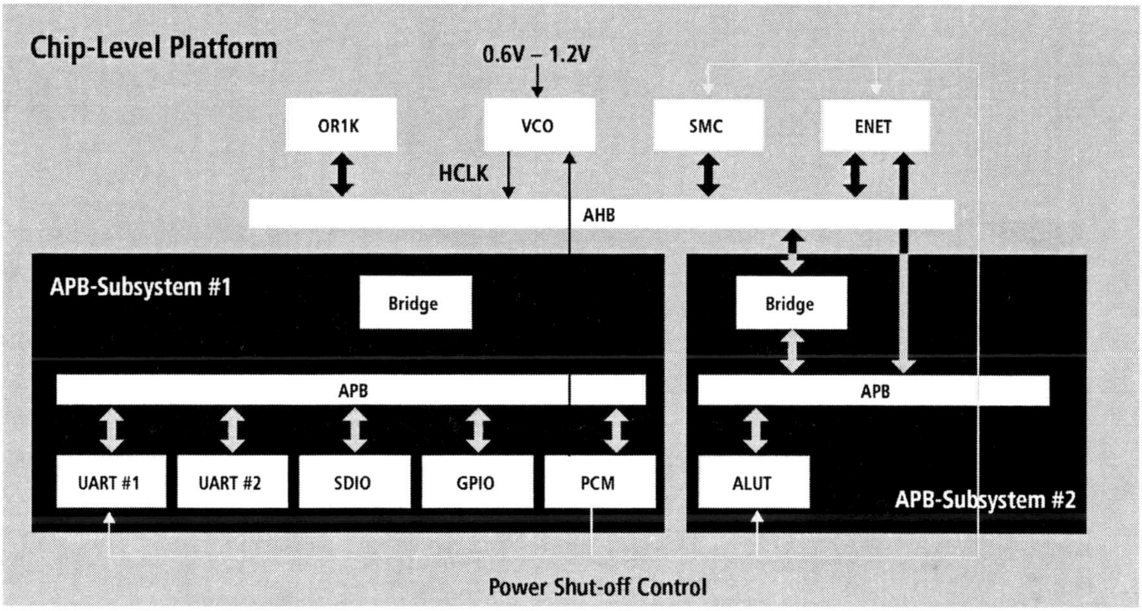

1.7 Reuse Isn't Just About Testbench Components

Universal Verification Components (UVCs) are reusable verification components that perform a vital role in accelerating testbench development through reuse. These components may entail complex developments, and commercial availability of UVCs for standard protocols provides substantial leverage for assembling complex testbenches. In all of this detail, engineers sometimes overlook the fact that many other pieces of the verification process may be reused in multiple dimensions.

UVCs provide project-to-project reuse at multiple levels. They also provide block-to-cluster reuse, which substantially boosts productivity when assembling a testbench for a hardware platform for which sub-component testbenches exist. Not only can the components be reused, but also sequence libraries, register definitions, and verification plans. The Incisive Verification Kit has comprehensive cluster-level environments that fully illustrate how to apply these reuse techniques.

Figure 1-8 on page 10 shows the architectural overview of a chip-level testbench and its corresponding chip-level vPlan, showing all of its verification plan reuse.

Figure 1-8 Architectural Overview of a Chip-Level Testbench and its Corresponding Chip-Level vPlan

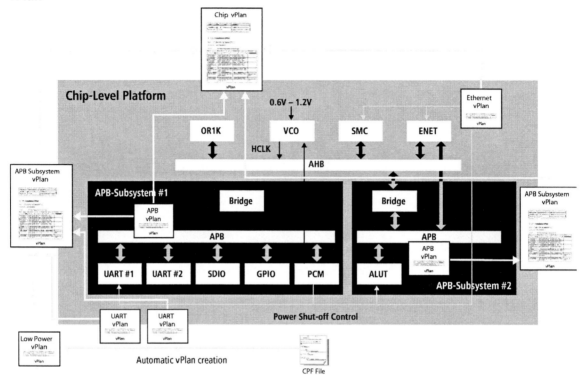

1.8 Does Speed Matter?

As a verification engineer, which of the following does your manager care more about: how fast your simulator runs or whether you achieve coverage closure on schedule? MDV enables sophisticated analysis of the coverage from the complete regression runs and allows you to correlate the results for a particular feature against the execution of the verification. In other words, MDV helps you answer the question, "Which simulations should I re-run in order to quickly test feature Y in 2 hours?"

Effective verification is not about running the most simulations in the least time; it is about running the most effective simulations as much of the time as possible. Simulations that increase coverage are the most valuable, as they are more likely to hit corner-case scenarios and therefore find bugs. MDV enables you to identify the random seeds and tests that contribute the most overall coverage or coverage for a particular feature or group of features. Running the whole regression suite more and more is like using a sledgehammer—eventually you will hit the target but it will take a lot of effort. Conversely, MDV is like putting a laser in the hands of the verification engineer; it can be focused and fired into specific areas where coverage holes exist. Figure 1-9 on page 11 shows the analysis of coverage contribution for each run of a regression against a specific vPlan feature.

Figure 1-9 Analysis of Coverage Contribution for Each Run of a Regression Against a Specific vPlan Feature

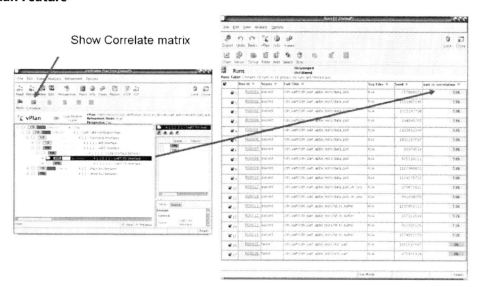

The Incisive Verification Kit provides workshops on how to use the analysis features of MDV so you can quickly get up to speed with this powerful capability.

1.9 What About Scalability?

One of the key lessons learned in the development of the UVM is the power of automatic test generation. By having constrained-random tests, each time a test scenario is run with a new seed, it is potentially going to find new bugs and increase coverage as it may traverse new states in the design. This scalability has massive potential when allied with infrastructure that supports vast processing farms. Huge numbers of concurrent simulations can be launched by one person, all potentially hunting bugs without the need for engineers to write hundreds and hundreds of directed test cases. This ability to scale the verification task is very compelling as it truly starts to automate the progress toward verification closure.

Figure 1-10 on page 12 shows an example of how MDV was applied to a real customer project. Using MDV, the customer not only reduced the project time frame from 12 months to 5 months but also found more bugs and used fewer engineers.

Figure 1-10 Example of MDV Applied to a Real Customer Project

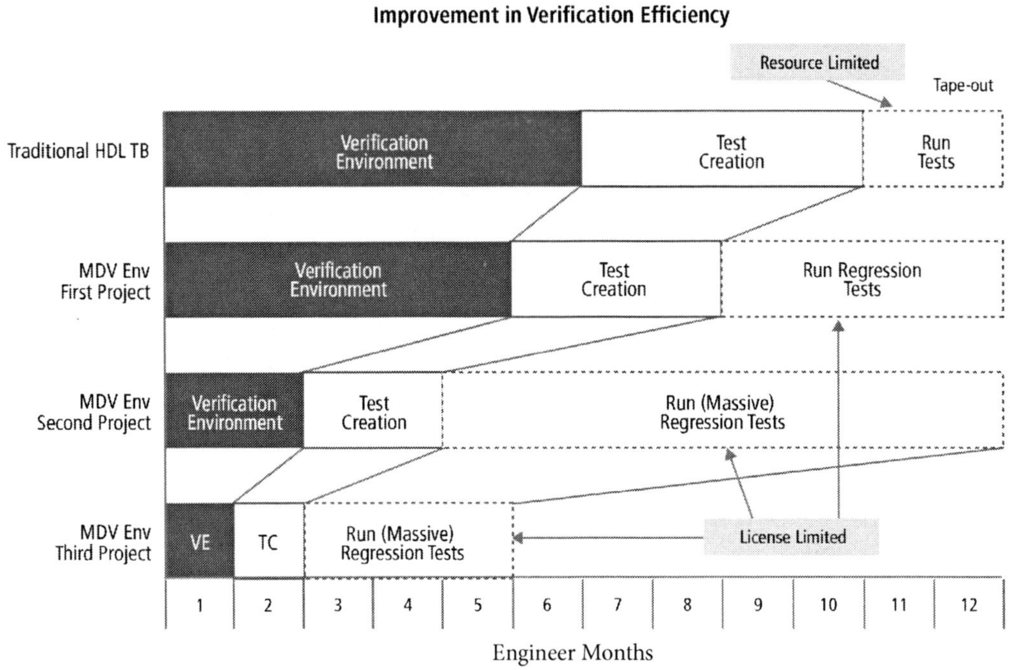

Improvement in Verification Efficiency

For formal verification a large set of formal properties traditionally constitute the verification task, and as the design grows and the suite of properties grow, so does the length of formal run time needed to complete the proof. The Cadence MDV approach enables a scalable "regression" approach to complete these large formal proofs through the use of compute farms. You can split a large set of assertions into a number of distributed smaller proofs, and thereby run the entire regression in a much shorter time. These types of tasks tend to be run repeatedly as chip sign off approaches, so the option of reducing run-time by a factor of 10 or 20 can make the difference between hitting a schedule or not.

1.10 Is Metric-Driven Verification Just for RTL Hardware?

Metric-driven verification is a structured and comprehensive approach to defining verification goals and measuring and analyzing progress toward those goals. It therefore is not confined explicitly to verifying hardware, but represents a by-product of the escalating cost of failure when developing and implementing complex silicon devices. And so it's perfectly reasonable to ask, "What systematic and comprehensive approaches do embedded software developers use to ensure adequate quality levels?" and "Could MDV be applied to embedded software?"

The answer to these questions is "yes." MDV techniques can easily be applied to embedded software, and the Incisive Verification Kit contains workshops and example flows of how to implement such technology. The techniques demonstrated show how to verify the software interface to the hardware, in a hardware-software co-verification flow.

One of the key points to keep in mind is the difference between validation and verification. Systems are typically designed and validated top-down, validation being the process to ensure that you are developing the right product. This is very much a software-oriented challenge, but still requires some form of hardware model— usually a functional virtual prototype built from transaction-level models (TLMs). Verification is usually performed as a bottom-up process that exhaustively proves the product has been developed right. Verification is normally performed to differing degrees on multiple abstraction levels. In other words, it is exhaustive verification at the block level, but perhaps integration and the programmer's view might be verified at the cluster or system levels.

Figure 1-11 shows the continuum that exists between hardware/software, verification, and validation, and the differing levels at which MDV might be applied.

Figure 1-11 The Continuum that Exists between Hardware/Software, Verification, and Validation

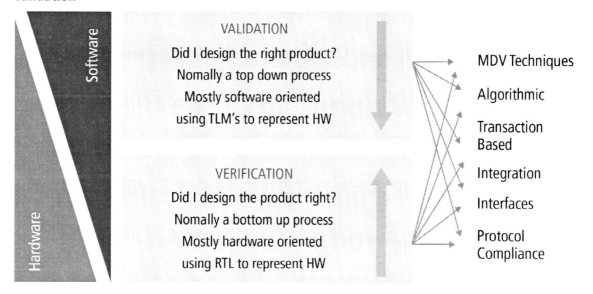

As embedded software content and complexity grows, there is more demand that deeply embedded products should work reliably while always powered on over longer and longer periods. The impact of software quality on product reliability will continue to rise in the same way that hardware quality has become mandatory for any new SoC design.

Deeply embedded software quality is a major concern for complex SoC products, especially as multi-core designs become more prevalent. Thus, the application of a rigorous, scalable quality measure such as MDV will become the norm.

1.11 How Do I Get Up to Speed with All this New Stuff?

So this new methodology and technology is great, you might say. But how on earth do you get my entire team up to speed in a realistic time frame, such that it doesn't impact my schedule?

The Incisive Verification Kit contains a large number of comprehensive hands-on workshops, including all the slides and lab instructions, so you can walk through a specific topic in your own time or perhaps ask a Cadence Application Engineer to run through it with you. The fact that these workshops are pre-packaged and delivered with the tools greatly enhances your ability to manage the ramp-up of teams on selected methodologies, all based on a realistic example SoC.

The current range of workshops delivered includes:

Assertion-based verification	• Introduction • Incisive Formal Verifier productivity flows • Incisive Formal Verifier connectivity flows
UVM – SystemVerilog	• Class-based SystemVerilog • Introduction to register modeling with reg_mem
UVM – Specman®/*e*	• Introduction to Specman • Introduction to *e* • IntelliGen debugging
UVM – mixed language (SystemVerilog and *e*)	• SystemVerilog over *e* • SystemVerilog with SystemC® TLMs
Metric-driven verification foundations	• Introduction • Planning • Infrastructure • Management • Automation
Metric-driven verification using verification IP	• Using multiple UVCs • Using the Compliance Management System • HW/SW co-verification

1.12 Summary

This chapter introduced the concept of metric-driven verification (MDV) and explained how it is a powerful layer of methodology that sits above the Universal Verification Methodology (UVM). The Incisive Verification Kit provides a comprehensive set of examples of how MDV can be applied across an entire range of verification technologies in a coherent and consistent way. Once applied, MDV puts powerful capabilities in the hands of engineers and managers. These MDV capabilities make the crucial difference; they improve verification effectiveness and, hence, managers and engineers alike can maximize their utilization of resources, use the breadth and depth of technology available, improve project visibility, and reduce the number of engineering man months.

2 UVM and Metric-Driven Verification for Mixed-Signal

2.1 Why Metric-Driven Verification for Analog?

In their 2006 paper titled "Verification of Complex Analog Integrated Circuits" Kundert et al. write:
Functional complexity in analog, mixed-signal, and RF (A/RF) designs is increasing dramatically. Today's simple A/RF functional block such as an RF receiver or power management unit can have hundreds to thousands of control bits. A/RF designs implement many modes of operation for different standards, power saving modes, and calibration. Increasingly, catastrophic failures in chips are due to functional bugs, and not due to missed performance specifications. Functionally verifying A/RF designs is a daunting task requiring a rigorous and systematic verification methodology. As occurred in digital design, analog verification is becoming a critical task that is distinct from design.[1]

Today, the challenges of analog verification have increased even further. Analog IP development is expensive, driving the reuse of IP into several SoCs. Each new integration brings about a new verification challenge because of configuration and connectivity changes. The verification of an analog block in the context of an SoC is tremendously difficult and time consuming. This highlights the need for thoroughly verified analog IP, as well as strong verification capabilities of the analog module in the SoC verification context.

Common methods of SoC-level analog verification include *black-box verification* of the analog portion and the *capture-and-replay approach*. The black-box approach involves integrating a highly abstracted model of the analog functionality, often just an interface with a loop-back for the digital signals. This prevents simulation speed degradation, but provides very little in terms of verification—it is often impossible to tell if the analog portion functions correctly, or whether it is exercised to a desired degree. The capture-and-replay approach extracts the boundary of the analog portion from the SoC simulation at some specific times (after initial configuration, for example). The waveform is then converted and replayed as input to an analog mixed-signal (AMS) simulation of the analog circuit. This method is better at exposing functional errors, but it is limited by the static nature of the analog-digital boundary captured. Key processes, such as calibration, require reactive feedback between the digital and analog portions, which is impossible to achieve using this method. Providing the exact timing and complex handshakes that may be required by the analog portion

---

1. *Verification of Complex Analog Integrated Circuits*, Ken Kundert & Henry Chang, In the proceedings of the IEEE 2006 Custom Integrated Circuits Conference (CICC)

makes the capture process complicated and error prone. Furthermore, the process is manual, lacking automated checking, severely limiting the number of cases that can be tested.

The deficiencies just described increase the risk involved in designing analog mixed-signal SoCs. Analog circuits may not be hooked up correctly, may not function correctly, and may not be driven as expected, while current verification methods may fail to detect the problems. Analog designs are often classified as "small A big D" or "small D big A" depending on the size of the analog content. Common wisdom suggests focusing on the "big" portion as the primary verification goal. In contrast, it is claimed that the critical aspect for verification is the level of interaction between the digital and analog portions. If the interaction is significant and complex, as it tends to be in modern circuits, the verification methodology must address the design as a whole, applying to both analog and digital portions with the same level of automation and rigor.

Digital verification engineering emerged in the last 20 years as an indispensable part of chip design. As complexities grow and productivity pressures rise, the expansion of verification engineering into the analog space in the short term is inevitable. This chapter outlines the methodology AMS verification engineers must follow to successfully address the challenge.

2.2 Approach and Scope

The UVM methodology has succeeded in tackling the hardest verification challenges in digital design. It is a metric-driven approach using coverage-directed random-stimulus generation, supporting multiple verification languages[1]. UVM promotes module-to-chip reuse and project-to-project reuse as means of maximizing development efficiency. It is the methodology of choice to extend for analog verification.

The extended methodology is named the Universal Verification Methodology–Mixed-Signal (UVM-MS). Methodology extensions include verification planning for analog blocks, analog signal generation, checking and assertion techniques for analog properties, and analyzing analog functional coverage. The methodology features abstract, high-level modeling of analog circuits using *real-number modeling* (RNM). Automation and management aspects include batch execution and regression environments, as well as progress tracking with respect to the verification plan.

Figure 2-1 on page 17 depicts a high-level view of the methodology, spanning both IP and SoC-level verification. At the IP level, a verification plan is created, the analog circuit is modeled, and a test environment is developed. For the best verification performance, the analog circuit should be modeled as an abstract RNM, though the methodology can also apply to AMS models and Spice netlists. In the figure, the flow of information is represented by the solid, angular arrows. The test plan, models and verification artifacts are reused at the SoC level, as indicated by the dashed, curved arrows.

---

1. Readers unfamiliar with UVM may benefit from reviewing the introduction at http://www.uvmworld.org/overview.php.

Figure 2-1   Applying UVM-MS at the IP and SoC Level

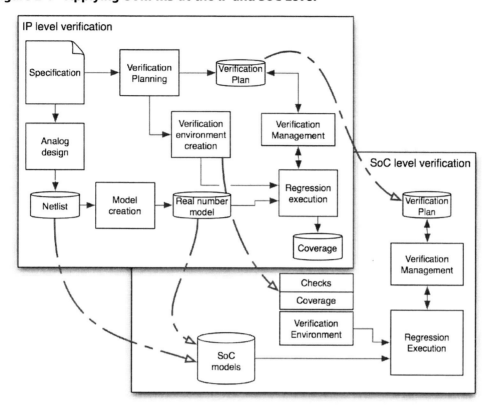

As part of the methodology, several teams are expected to collaborate in verifying mixed-signal designs. The analog architect and a team of analog designers are in charge of finalizing the specification, creating the design, and verifying its electrical properties. IP-level verification engineers implement a test environment and carry out metric-driven verification of the IP. The system-level integrator oversees the connection of the IP in the SoC. SoC verification engineers are in charge of integrating the IP-level test environment into the SoC level and the execution of the SoC-level testing. This separation of roles and responsibilities highlights the different areas of expertise required, in terms of domain knowledge, tools, languages, and methods.

The application of UVM-MS starts at the analog IP or module level to ensure correctness and robustness. This is done in parallel with the development of the analog circuit, and augments the verification done by the analog designer using SPICE-level analyses. The analog designers' work is mostly interactive and manual, and augmenting it with a metric-driven regression environment greatly improves quality and reduces risk, especially in view of rapid spec and design changes. Methodology and design tool evolution will enable the introduction of UVM-MS concepts into the analog designers' work processes, leveraging the interactive work as way of authoring checks and coverage.

The IP verification team with the help of the analog design team develops a verification plan. The plan needs to outline the properties to be verified, the testing scenarios, and coverage metrics that will ensure functional correctness. Subsequently, a UVM-MS verification environment is authored by the verification team. The verification environment is retarget-able to either the design netlist or an abstracted AMS or RN model. If abstract models are used, a batch mechanism to maintain model validity with respect to the netlist is included. The creation of a metric-driven verification environment at the IP level is done **in addition** to the traditional analog design flow, and is not intended to replace any portion of it. This additional investment is required to meet quality and risk profiles given current design complexity, specifically for:

- Verifying functionality and performance under all possible digital configurations, or a statistically meaningful portion thereof, if the number of combinations is too large.
- Verifying dynamic control scenarios, such as calibration for example, where digital controls are tied into a converging feedback loop.
- Verifying that control transitions, such as power mode and test mode switching, do not disrupt analog functionality. Here too, the sheer number of possible combinations typically requires randomly generated scenarios.

The large number of tests required to address the concerns listed above necessitates a metric-driven approach, featuring random generation. Directed testing cannot address these needs effectively. This IP-level process is depicted in Figure 2-2.

Figure 2-2   Verification Flow at the IP Level

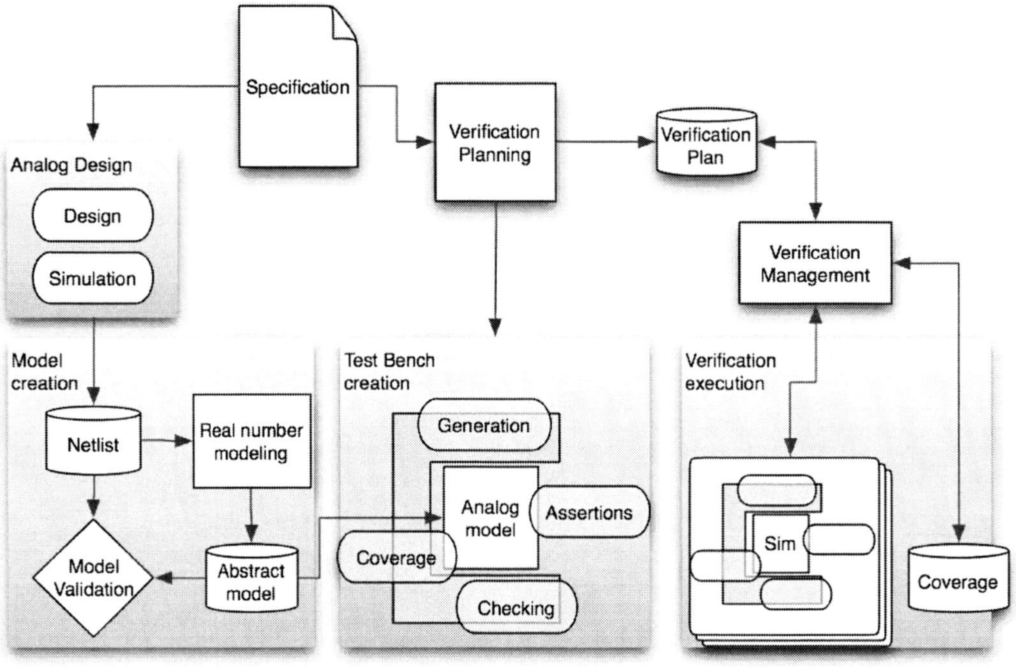

At the SoC level, the UVM-MS test plan is pulled into the SoC verification plan as a chapter. The verified abstract model of the IP is integrated in the SoC verification environment, enabling meaningful verification at reasonable simulation speeds. Components of the IP-level verification environment are reused, for the most part, minimizing the investment for creating the SoC verification environment and boosting confidence in its correctness. The reuse of verification artifacts during integration is depicted in Figure 2-3 on page 19.

The resulting SoC verification environment is highly capable, performing meaningful and thorough verification of analog functions and digital/analog interaction. This is achieved in spite of the knowledge gap that exists between the analog IP designer and the SoC integrator, thanks to the knowledge captured in the IP-level verification artifacts.

Figure 2-3   Verification at the SoC Level Reusing IP Verification Elements

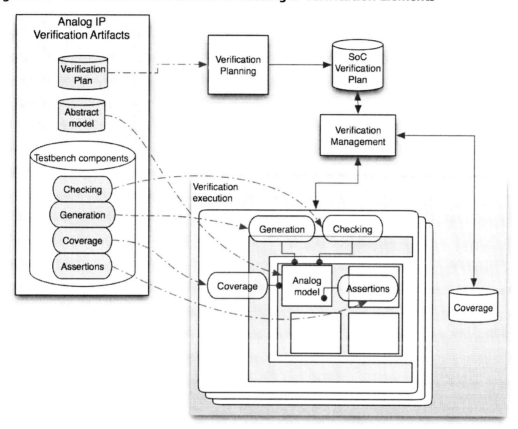

A Note About the Languages Used

The UVM-MS methodology has been developed using Verilog, *e*, and the Property Specification Language (PSL). The examples that follow are in these languages. This choice of languages is motivated by our wish to develop a best-in-class methodology that is not restricted by language limitations. The incorporation of VHDL/A, SystemVerilog, SystemVerilog Assertions (SVA) and possibly other languages will follow, reaching the language-agnostic stature of UVM.

2.3 Planning for Analog Verification

Metric-driven verification relies on a verification plan to be used as a starting basis. The plan lists all the features that need to be verified, **what** to check for and **how** to measure coverage. A typical plan describes test scenarios that would exercise each feature and important feature combinations.

Verification planning is best done using the Cadence Enterprise Planner (EP) tool. Using EP, it is possible to maintain a relation between an annotated spec, a section in the plan, and implementation code in the testbench. The EP flow recognizes the reality of partial specs, ever changing requirements and revisions of code, providing a mechanism for maintaining the three representations in sync.

2.3.1 Including Analog Properties

Verifying analog features often requires measuring continuous values, such as voltage or current at a certain node. Continuous (real) values can be sampled, but in order for them to make sense as coverage items they need to be quantized into bins. For example, a supply voltage may be classified as *nominal*, *low*, *high*, or *off*—creating a four-element coverage vector. More complex continuous properties, such as gain and signal-to-noise ratio can be computed based on several direct measurements, for example the signal amplitude at various locations in the data path. Such computed quantities need to be similarly quantized when captured as coverage items. Deciding what quantities to measure, either directly or indirectly, and how to quantize them needs to be part of the verification plan.

When considering analog circuits, the plan should include all properties of concern, including those that are not directly measured by a functional simulation (or transient analysis). For example, the frequency response of the circuit may be a property that requires verification. Rough estimates for some such properties can be computed based on a large number of functional simulations. Others may require SPICE-level analysis that could be automated as part of a regression system.

Including Analog Properties

Figure 2-4   Enterprise Planner Used to Capture Analog Property

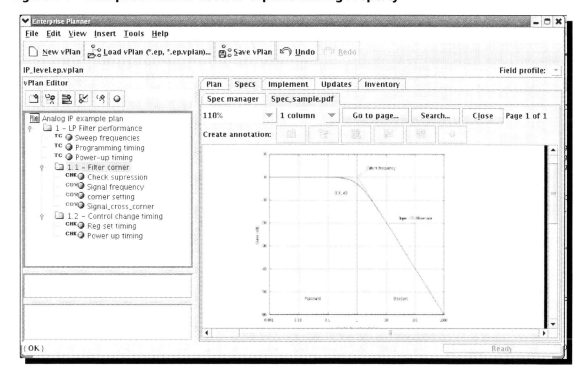

Analog property captured: attenuation of signal past corner frequency.

A programmable filter's coverage space is the cross of the programming values and the frequency range.

Special care should be taken to capture interactions between analog functions and digital controls. This includes control registers set during configuration and calibration, various operation modes, switching power modes, and so on. In the figure above, the low-pass filter is programmable. Verification obligations include verifying the frequency response (best done using AC analysis, but can also be checked using transient analysis with a number of frequencies). Another aspect is the delay between the control register setting and the filter response. This is harder to check, as one must not assume the response time is independent of the control value–that may be implementation dependent. Hence a test scenario including random control settings at random intervals, along with frequency variations is required. Collecting coverage of the parameters and checking the delay time meets the spec ensures correct functionality.

The contributors to an analog test plan should include the architect and analog design team. Information about corners and sweep settings (values included and excluded from sweeps) should be captured and documented in the plan even though they are not directly applicable in a digital setting. Capturing such information in the plan ensures that projects reusing the IP for integration and revisions have access to this information.

Figure 2-5 Plan Implementation Notes

Plan implementation notes include analog ranges and parameters, capturing designer knowledge about circuit sensitivity.

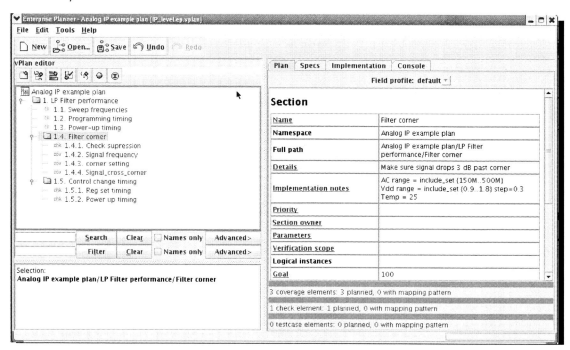

2.3.2   Verification Plan Structured for Reuse

The verification plan is used to guide the verification effort at the IP and SoC levels. This plan needs to be as complete as possible and has to be revised each time the spec, requirements, or implementation change. To facilitate this, the plan should be structured so that the mapping to specification sections is natural. The plan should be detailed enough such that each plan item (check, coverage, or scenario) maps directly to a single piece of code in the test environment. This often requires a revision once the testbench is implemented. An easy way to achieve that is by reading a coverage file into Enterprise Planner, and mapping each coverage point, check, and test case to a section in the plan.

Different integration targets will probably use the IP in different ways, often disabling features or locking them into a particular mode. Key values that are likely to change between projects could be defined as parameters, using the parameterized plan capability. A well structured plan will help identify which coverage points, checkers, and scenarios can be excluded and how to account for that when computing the overall coverage. Enterprise Planner provides convenient features for excluding sections from the overall coverage score as well as mapping logical scopes to physical scopes.

2.4 Constructing a UVM-MS Verification Environment

Creating a verification environment is a programming and modeling task carried out by a verification engineer, with reference to the verification plan. The environment needs to be capable of driving the circuit in all modes, conditions, and scenarios specified in the plan, checking all the properties specified, and measuring the requested coverage. The investment in a sophisticated verification environment is mitigated in two ways: leveraging pre-exiting universal verification components (UVCs), and re-using components of the verification environment during SoC integration and possibly project-to-project.

2.4.1 Analog Verification Blocks

UVM-MS features a few reusable verification blocks (mini-UVCs, in other words), that implement common tasks for analog verification. This section provides a high-level overview of some blocks that are used in subsequent examples. A complete list of available mini-UVCs and more detailed descriptions can be found in "UVM-MS Universal Verification Blocks" on page 72. Additional blocks will be added to the UVM-MS library as the methodology matures.

2.4.1.1 Signal Generator and Monitor

The *dms_wire* UVC features a programmable signal generator and monitor, which is used to drive and measure analog signals in a controlled manner. Requirements include:

- Driving a sine wave at a specified frequency, amplitude bias (DC level), and phase.
- Measuring the envelope of a signal, which is assumed to be periodic, within a time window. The time window is defined by a triggering event or a sequence item.
- Driving and monitoring configurations, which are controlled by dedicated sequence items and supporting easy integration into multi-channel test sequences. Controls can also be set by way of constraints for pre-run configurations.
- Checking coverage. The monitor features built-in coverage, customizable by extension, as well as a method port interface for custom coverage and checking.
- Resolving timing issues. Time resolution is explicitly controlled (the number of samples per cycle). Optionally, a continuous time signal source can be used for driving netlist models accurately.

It is possible to combine several signal generators to achieve complex input signals. Modulation and noise injection are common cases.

Figure 2-6 A Noisy Sine Wave Source Created by Composing Two Signal Sources

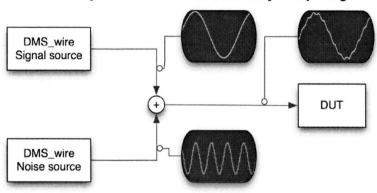

2.4.2 Analog Configuration Interface

Analog circuits often feature banks of control registers used to trim, program, or calibrate the circuit. These registers should be divided into distinct control interfaces according to their function, keeping in mind that any particular integration of the IP may treat such interfaces differently. For example, in one case an interface may be tied to a digital block driving it, while in a simpler instantiation that interface may be tied to constant values.

The simple *dms_reg* UVC example package provides the following functions:

- A simple macro construct for defining the interface, its control clock, and the set of registers included. Each register is specified with its size in bits and its reset value
- A sequence item interface that provides easy integration of interface setup commands (reset, write) into multi-channel test sequences.
- Value change notification through method ports, facilitating custom checking and coverage collection. Automatic coverage definition and collection is done upon value changes.

The *dms_reg* package is much simpler than the popular *vr_ad* register package. It offers very simple and limited functionality, and the code is not encrypted. Designs that make use of *vr_ad* at the SoC level may choose to use it instead of *dms_reg* at the IP level.

2.4.3 Threshold Crossing Monitor

Coverage and checking of aperiodic analog signals often require threshold crossing detection. Simpler cases use hard constants, for example, checking that a signal voltage is always below 0.9 Volts. More complex cases require a comparison between a signal and a reference. Because of model and circuit inaccuracies, this checking needs to allow for some tolerance, both in terms of the measured quantity and the response time. Consider for example the requirement: "*the regulator will maintain the output voltage at 95% +/-5% of the reference voltage, with a response time of less than t_{max}*".

Checking the relation between reference and signals shows that one signal is within voltage and time tolerances, while the other signal takes too long to settle after the step and shoots over the high margin.

Figure 2-7   Threshold Checking Example

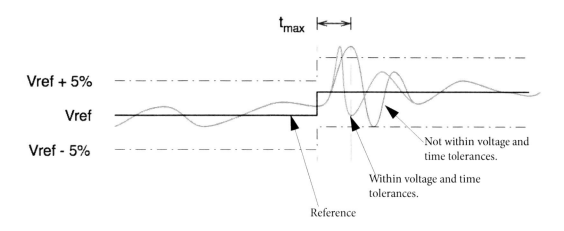

The *dms_threshold* block implements a programmable detector that is robust in face of short transients (spikes, glitches). It features:

- Transient suppression mechanism eliminating crossings (high-low-high and low-high-low) shorter than a set suppression time interval.
- Programmable high and low threshold and suppression time interval, set by way of constraints for pre-run configurations. The (nominal) threshold value is connected to a reference voltage signal.
- Event port outputs indicating definite threshold crossing (high-to-low, low-to-high and their union). Unfiltered high-crossed and low-crossed events can also be interfaced. This is for custom checking and coverage collection.
- Automatic coverage of crossings (low-to-high, high-to-low, both).

2.5   Architecture of a Sample Testbench

The following simple testbench architecture follows UVM guidelines, while enabling effective functional verification of an analog design. The verification components mentioned above provide an easy to control interface-to-analog portions. The top-level multi-channel sequence provides a single point of control for both analog and digital functionality, freeing the test writer from details concerning analog circuitry operation.

Figure 2-8 Architecture of a UVM-MS Environment

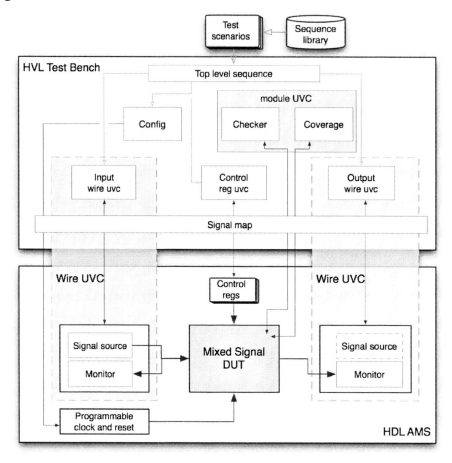

High-level functions are performed by the HVL portion, while high frequency functions are implemented in HDL

HVL = High-level verification language

HDL = Hardware description language

While much of the testbench functionality can be implemented using library blocks, some specific checkers and coverage monitors will need to be constructed. These are typically operate on sampled real numbers and are otherwise quite similar to monitors created for digital testbenches. These components should follow the UVM methodology guidelines. Some specifics are discussed in the following sections.

2.6 Collecting Coverage

Coverage collection involves capturing data at specified points in time. For analog coverage, deciding on **what** data to collect, **how** to categorize (continuous real value) data into bins and deciding on **when** to sample are the tasks at hand.

2.6.1   Direct and Computed Coverage Collection

Covered values can be voltage or current measured directly from circuit nodes. Other values may be computed by the testbench before sampling. Direct measurements are the simplest to implement—the covered node is sampled directly by the high-level testbench.

Example 2–1   *e* **Code Implementing a Directly Measured Real Coverage Item**

```
unit signal_map_u {
   -- Signal map connects the testbench to the design
   event rd_clk_fall is fall('~/tb_top/ck_rd')@sim;
   -- Real values must be accessed through ports
   real_vga_gain_db : in simple_port of real is instance;
   keep bind(real_vga_gain_db, external);
   keep real_vga_gain_db.hdl_path() == "~/tb_top/vga_ip/vga_gain_db";
};

unit gain_mon_u {
  smp    :signal_map_u; -- reference to signal map
  event gain_sample is @smp.rd_clk_fall;
  cover gain_sample is {
    item real_vga_gain_db : real = smp.real_vga_gain_db$ using ranges = {
      range([4.5..7.5],    "VAG-GAIN =  6dB +/- 1.5dB" );
      range([1.5..4.5],    "VAG-GAIN =  3dB +/- 1.5dB" );
range([0.0..1.5],    "VAG-GAIN =  0dB +/- 1.5dB" );
    }, illegal = (real_vga_gain_db > 7.5);
  };
};
```

Values that cannot be measured directly may be computed at the HDL testbench level for one of the following reasons:

- The quantity measured for coverage is not directly available. For instance, to compute the power of a node, one could create a real wire holding the multiplication of voltage and current. The result wire is then sampled for coverage.

- The source of the measured quantity differs depending on the DUT model being used. A value of interest may be available as an internal node in one model, a different node in another model, and a derived value in a third. In this case, replicating the value at a node in the HDL testbench allows the high-level testbench to deal with the different models uniformly.

- Coverage collected from within the analog model may be unavailable when running with a netlist model. Such selective coverage. should be defined under a *when* condition, so that it can be turned on and off as needed. For details, see "Trading Off Speed and Visibility" on page 30.

2.6.1.1   Methods for Timing the Collection of Coverage

Sampled signals may be highly dynamic, like a high-frequency data signal, or a slow changing one, such as a reference voltage. Slow changing data can be sampled using simple triggers. In contrast, naive sampling of dynamic signals will generate a broad spectrum of values depending on the exact sampling time. Consider the attempt to measure the amplitude of a sine wave —it cannot be done accurately using a simple trigger.

Trigger timing should take into account the event that causes a perturbation to the analog input or control, as well as the time it takes for the data path to respond and stabilize. Consider, for example, a digitally controlled voltage source. The change event is a write operation to the control register. Coverage should be sampled after the appropriate propagation delay, or response time, which is typically part of the circuit specification. The coverage event should be triggered by the write operation, delayed by the nominal response time.

Example 2–2   *e* **Code Implementing Delayed Coverage Collection**

```
unit signal_map_u {
   -- Signal map connects the testbench to the design
   event rd_clk_fall is fall('~/tb_top/ck_rd')@sim;
   -- following is the control register port
   vco_control : inout simple_port of int(bits:4) is instance;
   keep bind(vco_control, external);
   keep vco_control.hdl_path() == "~/tb_top/vco_ctrl[0:3]";
   -- following is the voltage regulator output
   real_vco_out : in simple_port of real is instance;
   keep bind(real_vco_out, external);
   keep real_vvco_out.hdl_path() == "~/tb_top/vco_out";
};

#define RESP_T 30;   -- response time is 30 ns, based on SN timescale
unit vco_mon_u {
   smp    :signal_map_u;
   event reg_written is change(smp.vco_control$)@smp.rd_clk_fall;
   event voltage_sample is {@reg_written; delay(RESP_T)}@smp.rd_clk_fall;
   -- voltage sample is triggered by register update and delayed by RESP_T
   cover voltage_sample is {
      -- cover the actual voltage output
      item real_vco_out : real = smp.real_vco_out$ using ranges = {
         range([1.1..2.5],    "High voltage > 1.1V" );
         range([0.9..1.1],    "Voltage in range 0.9 - 1.1V" );
         range([0.7..0.9],    "Voltage in range 0.7 - 0.9V" );
         range([0.0..0.7],    "Low voltage < 0.7V" );
      }, illegal = (real_vco_out > 2.5);
      -- add the control register setting to the cover group
      item vco_control : int(bits:4) = smp.vco_control$;
      -- cross coverage of output vs. control value
      cross vco_control, real_vco_out;
   };
};
```

In some cases it makes sense to sample the analog data path at given times, without an explicit reference to a control operation. An example would be a long training period of a modem—sampling the convergence along the way at points decided by the designer. In these cases, the coverage sampling event can be imperatively emitted (not triggered by a temporal expression). A sequence controlling the test can emit the

Direct and Computed Coverage Collection

event at the appropriate times. Note that in this case, the collection of coverage is dependent on the execution of the sequence—which might be left out during integration and re-use.

Coverage collection for rapidly changing signals requires an indirect measurement of the desired properties. The *dms_wire* UVC provides capabilities to monitor such signals and compute their amplitude, frequency, phase, and DC bias. These values cannot be sampled directly as mentioned above. A similar approach should be used to measure other properties of rapidly changing signals.

Example 2-3 *e* Code Collecting Coverage Properties of a Dynamically Changing Signal

```
unit tb_env_u {
    -- top-level environment, instantiating monitors
    sig_in  :dms_mswire_env is instance;
        keep sig_in.agent.active_passive == PASSIVE; -- monitor only
        keep sig_in.hdl_path() == "~/tb_top/sig_out";
        -- (other configuration constraints omitted for brevity)
    sig_mon :sig_mon_u is instance;
    -- connect the measurement output to the signal monitor method port
    keep bind(sig_in.agent.monitor.dms_mswire_transaction_complete,
              sig_mon.sig_in_meas);
};

unit signal_mon_u {
    -- method port connecting to DMS_Wire UVC monitor output
    -- presumably sig_in is a fast changing signal
    sig_in_meas : in method_port of
        dms_mswire_measurement_done_t is instance;
    -- internal variables for coverage
    !sig_ampl   :real;
    !sig_freq   :real;
    event in_meas_cover;

    -- Method port implementation
    sig_in_meas(rec :dms_mswire_measurement_t) is {
        message(LOW, "Input signal measured values:",
            "\n\tAmplitude =", rec.ampl,
            "\n\tBias      =", rec.bias,
            "\n\tPhase     =", rec.phase,
            "\n\tFrequency =", rec.freq);
        sig_ampl = rec.ample;
        sig_freq = rec.freq;
        emit in_meas_cover;
    };

    -- Coverage definition
    cover in_meas_cover {
        item freq : real = sig_freq using ranges = {
            range([7e8..1e9],    "700M-1GHz" );
            range([5e8..7e8],    "500M-700MHz" );
            range([2e8..5e8],    "200M-500MHz" );
```

```
                range([1e8..2e8],      "100M-200MHz" );
        }, illegal = (freq < 1e8);
        item ampl : real = sig_ampl using ranges = {
                range([5e-3..20e-3],   "5-20mV" );
                range([1e-3..5e-3],    "1-5mV" );
                range([0.0..1e-3],     "Below 1mV" );
        }, illegal = (ampl > 20e-3);
    };
};
```

2.6.2 Deciding on Coverage Ranges

The simulated model may be continuous or event driven, depending on the simulation engine, but the sampled coverage is always quantized. Coverage results are presented in the resolution of coverage ranges as defined in the cover statement. It is therefore important to consider the ranges associated with analog coverage.

Primarily, one should take into account any ranges explicitly mentioned in the spec for the quantity being measured. Key ranges may be mentioned in the verification plan as well. Next, special ranges should distinguish *reset*, *power-down*, and *inactive* values (the value expected when there is no input signal for example). Normal operation values should be split if there is a special range of importance, below or above a corner of a filter, for example. Finally, unexpected and erroneous ranges should be considered. One advantage of using ranges is the ability to tag a specific range with a name, creating a classification of measured results. Tagging a range with "Unexpected low gain" is much more meaningful than "1-3dB", especially for verification engineers dealing with integration at an SoC level, who may not be familiar with the design. Having thoughtful ranges and name tags is a good way of transferring design knowledge along.

In addition to defining coverage items with well-thought-out range specifications, one should also consider defining cross coverage. A good starting point is the cross of control register settings with the analog coverage they should affect. This provides a more meaningful picture of the functionality covered. Cross coverage of analog cover items provides a way of measuring corner cases.

2.6.3 Trading Off Speed and Visibility

Coverage is a powerful tool for guiding verification. At a block-level it may make sense to have very detailed coverage. However, moving to the SoC-level the impact of coverage collection on run time may become more significant. When designing and implementing coverage one has to plan for trading off speed and coverage resolution. It is recommend that you separate interface coverage, sometimes called *black-box* coverage from internal or *white-box* coverage. During integration, the emphasis is primarily on black-box coverage.

The most structured way to throttle coverage is to define a global model in the test environment, determining coverage resolution. In a black-box mode, high-level events visible on the interface and key analog metrics should be collected. Any complex computations or high frequency sampling should be restricted to one or more detailed coverage modes. Coverage implementation should make use of these global modes to allow selective switching of coverage. A simple interface at the top level should allow the SoC integrator to control coverage collection for the IP without having deep knowledge of the IP design or test environment.

Example 2-4 *e* Code Using when Construct to Declare Different Testbench Configurations

Note Coverage of nodes internal to the DUT is dependent upon configuration.

```
-- declare a type to classify testbench configuration
type tb_kind_t :[FULL_VISIBILITY, INTF_ONLY_VISIBILITY];

uint tb_env_u {
    kind : tb_kind_t;   -- top-level configuration control
    keep soft kind == FULL_VISIBILITY; -- use full visibility by default
    smp     :signal_map_u is instance;
    keep smp.hdl_path() == "~/tb_top";
    keep smp.kind == kind;
    gain_mon  :gain_mon_u is instance;
    keep gain_mon.kind == kind;
};

unit signal_map_u {
    -- Signal map connects the testbench to the design
    kind : tb_kind_t;
    keep soft kind == FULL_VISIBILITY; -- dive in by default
    -- testbench-level signals are accessible under all configurations
    clk_port          : input simple_port of bit is instance;
    keep clk_port.hdl_path() == "ck_rd"
    event rd_clk_fall is fall(clk_port$)@sim;
    vga_gain_control : inout simple_port of int (bits:3) is instance;
    keep bind(vga_gain_contol, external);
    keep vga_gain_control.hdl_path() == "vga_gain_ctrl[0:2]";
    -- nodes internal to the DUT are conditional
    when FULL_VISIBILITY signal_map_u {
      real_vga_gain_db : in simple_port of real is instance;
      keep bind(real_vga_gain_db, external);
      keep real_vga_gain_db.hdl_path() == "vga_ip/vga_gain_db";
    };
};

unit gain_mon_u {
  smp     :signal_map_u; -- reference to signal map
  kind :tb_kind_t;
  keep soft kind == FULL_VISIBILITY;

  event gain_sample is @smp.rd_clk_fall;
  cover gain_sample is {
    -- this part of the cover group is sampled unconditionally
    item vga_gain_control : int(bits:3) = smp.vga_gain_control$;
  };
  when FULL_VISIBILITY gain_mon_u {
    -- access to DUT nodes is otherwise restricted
    cover gain_sample is also {
      item real_vga_gain_db : real = smp.real_vga_gain_db$ using ranges = {
```

```
                range([4.5..7.5],    "VAG-GAIN =  6dB +/- 1.5dB" );
                range([1.5..4.5],    "VAG-GAIN =  3dB +/- 1.5dB" );
                range([0.0..1.5],    "VAG-GAIN =  0dB +/- 1.5dB" );
            }, illegal = (real_vga_gain_db > 7.5);
        };
    };
};
```

2.7 Generating Inputs

Analog IP may have both digital and analog inputs, in addition to numerous configurations and other settings. The general approach for generation is similar to UVM.

2.7.1 Dealing with Configurations and Settings

The test environment may be able to control various configurations and settings dynamically, while others may be hard-wired, requiring a re-compilation if altered. Dynamically changeable configurations should be modeled as fields in a configuration object. Constraints are used to restrict the legal ranges and to select specific configurations in specific test files. The configuration object should be generated once upon start-up and a configuration method should be called to drive the generated values to the controls in the device under test (DUT) and the HDL test environment.

In some cases, the IP designer might have used `define Verilog macros to set some parameters. The macros should be read into the testbench and the values used to constrain and drive the parameters. This retains the designer's intentions, while providing a clear path of control (all values are driven by the testbench, eliminating multiple controls through macros scattered throughout the HDL code).

Example 2-5   *e* **Code for Generating and Applying Configuration**

```
verilog import design_params.v;   -- read in `defined macros
struct tb_config_s {
    smp        : tb_smp_u;    -- reference to signal map (not included here)
    resolution :real;         -- determines the tb sampling frequency
    keep soft resolution == 0.2; -- ns per sample
    clk_delay  :uint;         -- this value controls the TB generated clock
    keep clk_delay in [10..100];
    vdd_v      :real;    -- vdd_v is read from an HDL `define macro
    configure() is {
        -- this method is called to apply the configuration
        smp.tb_resolution$ = resolution;
        smp.clk_delay$ = clk_delay;
        vdd_v = `HDL_V_VOLTAGE;   -- defined in design_params.v
    };
};

unit tb_env_u {
    smp     :tb_smp_u is instance;
    config  :tb_config_s;
```

```
        keep config.smp == smp;
        -- config will be generated upon startup
        run() is first {
            -- apply configuration prior to starting the run
            config.configure();
        };
};
```

2.7.2 Generating and Driving Digital Control

Digital control is generated and driven according to the UVM. Each interface has a bus functional model (BFM) that receives interface-specific transactions and uses them to drive digital values. Transactions are driven into the BFM by a sequence driver. The sequence driver is controlled by a sequence. Each interface may have a specific sequence library containing common building blocks for sequences. For example a control register interface may define a transaction object containing an address field, a data field, and a flag for *read* or *write* operation. The BFM can translate each transaction object to the appropriate sequence of operations needed to perform the read or write requested. The sequence library may contain sequence fragments for more complex operations, like reading a few registers, or resetting all.

This arrangement results in multiple levels of generation, reflecting the nesting of transactions within sequence hierarchies. Constraints may be specified at the sequence item level as well as the transaction type definition. The transaction should be defined such that only desired values are generated (for example, legal values and erroneous values used for stressing the interface). The transaction should include soft constraints leading to the generation of legal input.

At each sequence level, constraints should be as broad (non-limiting) as possible. This will ensure the testbench is not avoiding some combinations unintentionally. It will also make the sequence more reusable.

The code example below contains a skeletal setup used to control a power-up/down interface.

Figure 2-9 on page 34 depicts the architecture of the controller, and Example 2–6 on page 34 lists the actual code. The testbench interface to this module is a sequence driver, accepting sequence items specific to this interface. Thus the interface can be controlled by the top-level multi-channel sequence. The BFM unit of the power controller executes a specific hard-coded procedure, presumably taken from the specifications, to power the circuit up or down.

Figure 2-9 Architecture of a Power Controller Driven by Sequence

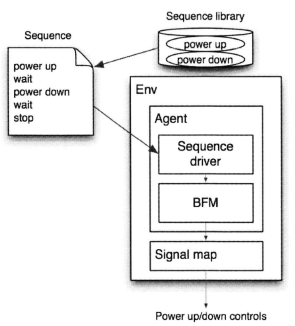

Example 2–6   *e* Code Example—Power Control Interface

```
struct power_control_transaction like any_sequence_item {
    kind :[PWR_UP, PWR_DN];
    pu_delay :int;
    keep pu_delay in [10..200];
};

-- define the sequence - this creates the driver, sequence item etc.
sequence power_control_sequence using
    item = power_control_transaction,
    created_driver = power_control_driver_u,
    created_kind = power_control_sequence_kind_t;

extend power_control_driver_u {
    keep soft gen_and_start_main == FALSE; -- prevent bogus item on startup
};
-- The BFM portion
unit power_control_bfm_u {
    p_smp     :tb_smp_u; -- reference to signal map (not included here)
    driver    :power_control_driver_u is instance;
    trans     :power_control_transaction;

    event clk is @p_smp.rd_clk; -- assuming its available when power is off!
    on clk {
        emit driver.clock;
```

```
    };
    
    run() is also {
        reset();
        start get_and_drive();
    };
    
    reset() is {
        -- bring circuit power controls to reset state
        p_smp.powerup$ = 0;
        p_smp.power_clamp$ = 1;
        p_smp.power_regulated$ = 1;
    };
    
    get_and_drive() @clk is {
        while TRUE {
            trans = driver.get_next_item();
            -- NOTE: use start here to allow concurrency
            -- don't want the calling sequence to get stuck while
            -- powering up...
            start drive_transaction(trans);
            emit driver.item_done;
        };
    };
    
    drive_transaction(trans :power_control_transaction) @clk is {
        -- some delays here are constant, while others are
        -- created as random values in the transaction
        message(HIGH,"Driving power control transaction kind=",
            trans.kind);
        if trans.kind == PWR_UP then {
            p_smp.powerup$ = 1;
            wait delay(trans.pu_delay);
            p_smp.power_clamp$ = 0;
            wait delay(250);
            p_smp.power_regulated$ = 0;
        } else { -- power down presumably
            p_smp.power_regulated$ = 1;
            p_smp.power_clamp$ = 1;
            wait delay(trans.pu_delay);
            p_smp.powerup$ = 0;
        };
    };
};
```

2.7.2.1 Generating and Driving High-Frequency Analog Signals

High-frequency analog signals are not generated directly. Rather, the properties of the signal being driven are controlled by digital values, which are generated *in turn*. For example, the frequency of a signal may be determined by a real value, which is generated in turn by the testbench. This holds even for signals that should have some randomness, like noise, because the overhead of generating numbers makes a direct approach impractical.

Thus, analog input that is not simply DC, or a slow changing signal like a ramp, needs to be a periodic waveform like a sine wave or a sawtooth, or some composition of such sources. The *dms_wire* UVC features a simple sine wave generator, which is easily changeable to generate other periodic signals. The signal generator is controlled by four control values determining the frequency, phase, amplitude, and DC bias of the generated signal. The generator is configurable to drive either a time quantized real value or a continuous electrical signal as needed.

The signal generator's control is structured according to the UVM, just like digital control interfaces. A transaction object contains fields for all the control parameters. There is a BFM which converts the transaction to a setting for the signal generator. Sequence items provide simple control interface for the test writer, making it easy to control the generated signal on the fly. The *dms_wire* UVC is available in the UVM-MS library.

The example below shows a simple sequence controlling a signal generator. A more elaborate example is included in the next section.

Example 2-7 *e* **Test File Defining a Sequence that Sets Up Two Signal Generators and Measures a Resulting Signal**

```
-- This sequence is defined in a test file
extend MAIN dms_tb_sequence {
    body() @driver.clock is only {
        out("Signal source 1 set up");
        do DRIVE_SEQ dms_sequence on driver.osc1 keeping {
            .ampl == 20.0;
            .bias == -5.5;
            .freq == 10.0e6;
            .phase == 0.0;
        };
        out("Signal source 2 set up");
        do DRIVE_SEQ dms_sequence on driver.osc2 keeping {
            .ampl == 0.5;
            .freq == 1.0e8;
            .phase == 90.0;
        };
        out("Measure");
        do MEASURE_SEQ dms_sequence on driver.mon keeping {
            .delay == 400.0;
            .duration == 400;
        };
        wait [100];
        out("Done");
        stop_run();
    };
};
```

2.7.2.2 Top-Level Multi-Channel Sequence

To maintain full control of interaction and timing across multiple interfaces, a single point of control is required. This is supported in the UVM by the top-level sequence (sometimes called *multi-channel sequence* or *virtual sequence*). A typical sequence has a single sequence driver, which receives and drives each sequence item. The top-level sequence holds references to multiple sequence drivers, one for each active interface. The top-level sequence can generate and pass sequence items to any appropriate interface. Thus, the top-level sequence is able to orchestrate activity across all interfaces.

The UVM-MS follows the exact same structure. The analog nature of the design is only reflected in some of the sequence items being used—for instance, the ones controlling analog signal sources. The overall ease of use for test writers is maintained and so is the modularity and reusability of tests. Test files are most typically just variations on the top-level sequence.

The following example has an analog filter circuit embedded in a testbench containing a noisy signal source and a checker. The signal source is constructed by summing together a low frequency "data" sine wave and a high frequency "noise" sine wave. The testbench is controlled by a multi-channel sequence that drives the three components in an orchestrated manner.

Figure 2-10   An Example of a Top-Level Sequence Controlling Three UVCs

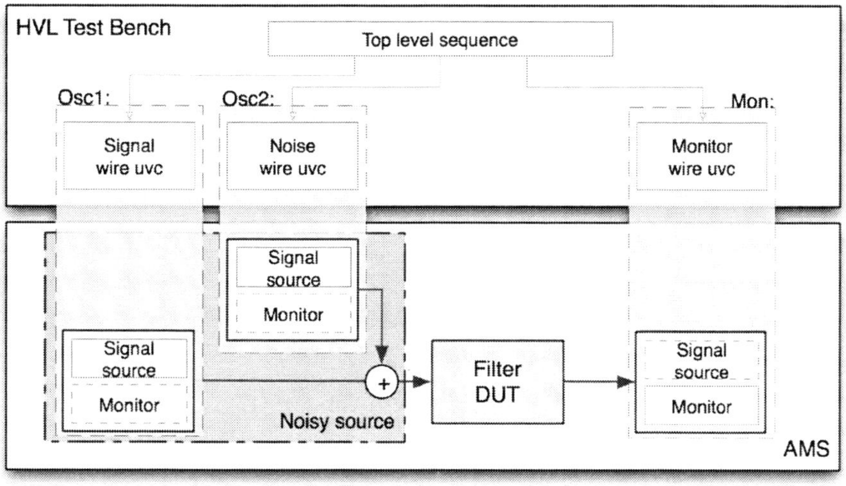

Example 2–8   *e* Code Implementing a Top-Level Sequence Skeleton

Note   A test file such as the one in the example code can be loaded on top to program a specific sequence.

```
unit dms_tb_env_u like any_env {
    -- instantiate the UVCs
    osc1 :dms_mswire_env is instance;
        keep osc1.agent.active_passive == ACTIVE; -- make it drive
        keep osc1.hdl_path() == "~/dms_top/sig_gen1";
    osc2 :dms_mswire_env is instance;
        keep osc2.agent.active_passive == ACTIVE;
```

```
        keep osc2.hdl_path() == "~/dms_top/sig_gen2";
    mon    :dms_mswire_env is instance;
        keep mon.agent.active_passive == PASSIVE; -- monitor only
        keep mon.monitor_mode == TRANS_DRIVEN; -- controlled by transaction
        keep mon.hdl_path() == "~/dms_top/sig_mon";
    -- instantiate the virtual sequence driver
    driver :dms_tb_sequence_driver is instance;
        keep driver.osc1 == osc1.agent.driver;
        keep driver.osc2 == osc2.agent.driver;
        keep driver.mon  == mon.agent.driver;
};

sequence dms_tb_sequence; -- defines the top-level sequence

extend dms_tb_sequence {
    -- add a field of type sequence to support "do" actions
    !dms_sequence      :dms_mswire_sequence;
};

-- create references to the drivers of all channels
extend dms_tb_sequence_driver {
    osc1 :dms_mswire_driver_u;
    osc2 :dms_mswire_driver_u;
    mon  :dms_mswire_driver_u;
    -- generic clock for virtual sequence driver
    event clock is only @sys.any;
};

extend sys {
    -- instantiate the testbench
    tb_env :dms_tb_env_u is instance;
};
```

2.8 Checking Analog Functionality

The UVM-MS approach to checking is possibly its biggest departure from current practices in analog design. When using metric-driven verification there is simply no escape from automatic checking, while manual inspection of waveforms is still a much-used method for checking analog correctness. The challenge of automating analog checking is not to replace manual inspection completely, but rather to provide a reasonably good detection of errors for the multitude of batch runs. Deciding which functions and features to check for is a major decision during the verification planning phase. Some typical checking patterns are described below.

2.8.1 Comparing Two Values

Often the checked property involves a relationship between two signals—the input and output of an amplifier for instance. The compared property has to be determined and for an amplifier this will probably be the amplitude of each signal. Other properties may include the phase shift or DC bias.

The properties of each signal need to be measured independently, possibly by using the *dms_wire* monitor. The checker will compare the measured results with the expected relationship.

Timing the check requires determining a triggering event to start the measurement on both signals. In case there is an expected delay between input and output, the measurements need to be delayed accordingly. The measurement occurs over a period of time, at the end of which, each monitor provides its output, typically as a call to a method port. The final checker code is triggered by the returned measurements.

Figure 2-11 A Checker Comparing DUT's Input and Output to Compute Gain

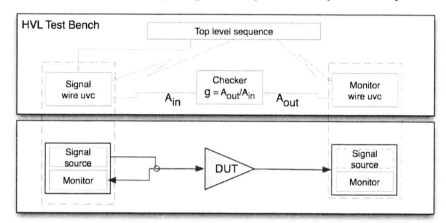

A checker comparing the DUT's input and output to compute gain. The measurement is triggered by the checker.

Figure 2-12 Timing Sequence for the Gain Checker

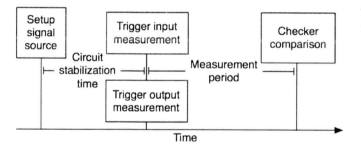

The circuit needs to hold stable for a measurement period.

Example 2-9   *e* Code Example of Gain Checker

```
unit gain_checker_u {
    -- method ports connecting to monitors
    amp_in_meas  : in method_port of
        dms_mswire_measurement_done_t is instance;
    amp_out_meas : in method_port of
        dms_mswire_measurement_done_t is instance;
    -- internal variables for checking
    !amp_in_ampl    :real;
    !amp_out_ampl   :real;
```

Advanced Verification Topics

```
          event in_meas;
          event out_meas;
          -- check_gain triggers when both measurements are in - temporal "and"
          -- means simultaneous occurrence - no delay in this example
          event check_gain is @in_meas and @out_meas;

          amp_in_meas(rec :dms_mswire_measurement_t) is {
              amp_in_ampl = rec.ampl;
              emit in_meas;
          };
          amp_out_meas(rec :dms_mswire_measurement_t) is {
              amp_out_ampl = rec.ampl;
              emit out_meas;
          };
          -- The actual gain checker code
          on check_gain {
              var gain :real = (amp_in_ampl == 0 3 0 :
                                amp_out_ampl / amp_in_ampl);
              var measured_gain_db  :real = (gain == 0 3 0 : log10(gain) * 20);
              -- Spec requires minimum gain of 15 dB +/- 1.5
              check that (measured_gain_db >= 16.5)
              else dut_error("Gain checker - measured gain ", gain, " ( ",
                             measured_gain_db," dB) ",
                             " is below minimal gain of 15 dB +/- 1.5");
          };
     };
```

2.8.2 Triggering a Check on Control Changes

Carefully timing a check in a specific test may be good enough for a feature test, but having the check trigger automatically is much more powerful. Automatic activation of tests ensures the checked condition is monitored continuously, even when that feature is not targeted. This is especially important when the checker is integrated in a bigger environment where tight control over input timing may not be possible.

To determine the activation timing of a check, consider which controls affect the checked condition. These controls are likely to be register settings and external interfaces. Each time any of these controls change, the check should be triggered. The following timing diagram illustrates this concept.

Figure 2-13 Triggering Checks Automatically Upon Control Value Change

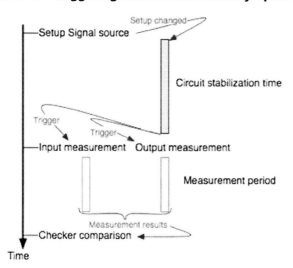

Triggering checks automatically upon control value change. In this case, measurement is delayed to allow the circuit to stabilize after the change.

It is important to note that measuring the properties of a dynamic analog signal takes some time, called here the *measurement period*. During this period the measured signal is over-sampled and properties, such as its frequency and amplitude, are calculated. The minimal measurement period depends on the signal frequency, and should be no shorter than five cycles to ensure reliable results.

Because checking analog properties may involve measurements that are performed over some period of time, one has to protect against tightly staggered activations of the checker. A simple *busy bit pattern* is sufficient in most cases, as there is typically no concurrency in the checker control code.

When control registers are affecting the checker, there is often a change event that can be used to trigger the checking. In the case of the *dms_reg* UVC package, a "change occurred" event port is available per the interface. The example below illustrates how such an event port can be used to trigger the monitor measurement, which in turn will activate the checker code. The result is a fully automatically timed checker. (The checker code is the same as the one in Example 2-9 on page 39).

Example 2-10 *e* Code Activating Amplitude Measurement upon Change in Control Registers

```
unit amp_tb_env_u {
    -- Source and Monitor for amp_in
    amp_in   :dms_mswire_env is instance;
        keep amp_in.agent.active_passive == ACTIVE;
        keep amp_in.hdl_path() == "~/tb_top/sig_gen";
        -- monitor portion is on as well:
        keep amp_in.monitor_mode == EVENT_DRIVEN;
    -- Monitor observing amp_out
    amp_out  :dms_mswire_env is instance;
        keep amp_out.agent.active_passive == PASSIVE;
        keep amp_out.monitor_mode == EVENT_DRIVEN;
        keep amp_out.hdl_path() == "~/tb_top/sig_mon";
```

UVM and Metric-Driven Verification for Mixed-Signal

```
        -- Control interface
        amp_vga     :dms_reg_amp_vga_env is instance;
            keep amp_vga.agent.active_passive == ACTIVE;
            keep amp_vga.hdl_path() == "~/tb_top";
            -- changes detected in control reg values trigger measurements
            keep bind(amp_vga.smp.change_occurred,amp_in.smp.do_sample);
            keep bind(amp_vga.smp.change_occurred,amp_out.smp.do_sample);
    };
```

In case the measurement of the input and output are not well synchronized, the checker should latch measured values and requires a little more sophistication. An example is presented below.

Example 2–11 This Checker Doesn't Assume Input and Output Samples are Synchronized

```
    unit gain_checker_u {
       -- method ports connecting to monitors
       amp_in_meas : in method_port of dms_mswire_measurement_done_t is instance;
       amp_out_meas : in method_port of dms_mswire_measurement_done_t is instance;
        -- internal variables for checking
        !amp_in_ampl    :real;
        !amp_out_ampl   :real;
        !in_amp_valid;     -- TRUE as long as the sampled data is valid
        !out_amp_valid;    -- TRUE as long as the sampled data is valid

        amp_in_meas(rec :dms_mswire_measurement_t) is {
            amp_in_ampl = rec.ampl;
            in_amp_valid = TRUE;
            check_gain();
        };
        amp_out_meas(rec :dms_mswire_measurement_t) is {
            amp_out_ampl = rec.ampl;
            out_amp_valid = TRUE;
            check_gain();
        };
        -- The actual gain checker is now a method
        check_gain() is {
            if not(in_amp_valid and out_amp_valid) then {
                return;    -- Both samples are not ready yet
            };
            in_sample_valid = FALSE;
            out_sample_valid = FALSE;
            -- perform the check, because both samples are in
            var gain :real = (amp_in_ampl == 0 ? 0 : amp_out_ampl / amp_in_ampl);
            var measured_gain_db  :real = (gain == 0 ? 0 : log10(gain) * 20);
            -- Spec requires minimum gain of 15 dB +/- 1.5
            check that (measured_gain_db >= 16.5)
            else dut_error("Gain checker - measured gain ", gain, " ( ",
                    measured_gain_db," dB) ",
                    " is below minimal gain of 15 dB +/- 1.5");
        };
    };
```

2.8.3 Measuring Signal Timing

A signal may be expected to match a certain shape, for instance reaching a certain voltage in a given time window after power-up. Such checks can be implemented by measuring the signal in specific time windows.

A typical timing checker is triggered by the initial event, the power-up event in this example. Following the specified delay, the signal is measured and compared against the specified value. Tolerances may be used when comparing the measured value. If needed, additional delay and measure steps can be added to ensure the signal matches the specified timing. Each measurement point should have its own check so that error messages can be specific and checks can be controlled independently.

Figure 2-14 Checking for Signal Timing

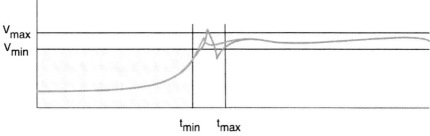

The checking window is bounded by minimal and maximal time and voltage. Two check obligations are (1) signal is below threshold until tmin; and (2) signal is within Vmin and Vmax at time tmax.

Simple timing checks verify a value at a point in time. Such checkers will not be able to distinguish between the two waveforms in the figure above. If such resolution is needed, a more sophisticated threshold checker has to be used. That topic is covered in the next section.

Dynamic signals cannot be measured by direct sampling. The *dms_wire* monitor can be used in these cases. Since measurement takes time, be sure to trigger the measurement without introducing extra delay into the checking sequence. The checker sampling period must be much shorter than the time windows being checked in order to approximate the desired zero-time check at the end of the time window, as can be seen in the following figure.

Figure 2-15 Timing a Measurement of a Dynamic Signal for Checking DC Bias at Time t-max

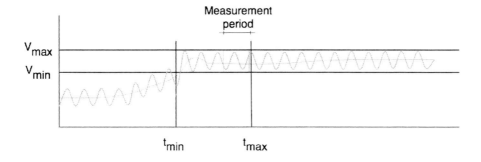

Example 2-12 *e* Code for Two Simple Checkers Verifying Voltage/Timing

The first checker verifies voltage at given time, the second verifies timing at certain voltage.

```
unit pwr_tb_checker_u {
    p_smp         :tb_smp_u;    -- reference to signal map
    pwr_v_tolerance  :real;      -- defines voltage tolerance for checker
    keep pwr_v_tolerance == 20.0; -- mV
    pwr_sample_delay :uint;     -- sampling for slow changing power signal
    keep pwr_sample_delay == 10;  -- ns
    !pwr_rise_time   :time;
    event powerup_rise is rise(p_smp.powerup$)@sim;
    on powerup_rise {
        power_rise_time = sys.time;
        start power_up_checker1();
        start power_up_checker2();
    };

    power_up_checker1()@sys.any is {
        -- A time consuming method that measures the regulated power
        -- voltage 100 ns after invocation and compares
        -- with spec'ed values
        var pwr_meas_val   :real;
        wait delay(150);
        pwr_meas_val = p_smp.reg_power_v$; -- port reflecting voltage
        check that abs(pwr_meas_val - 750) < pwr_v_tolerance
        else dut_error("Power up - regulated power failed to reach " ,
               "spec voltage of 750 mV +/-", pwr_v_tolerance,
               " 150 ns after power-up - measured ",
               pwr_meas_val);
    };

    power_up_checker2()@sys.any is {
        -- A checker method that samples the power signal and ensures
        -- it remains below spec'ed value for the first 100 ns
        var pwr_meas_val   :real;
        while (sys.time < pwr_rise_time + 100) {
            wait delay(pwr_sample_delay);
            pwr_meas_val = p_smp.reg_power_v$;
            check that pwr_meas_val < 650+pwr_v_tolerance else
                dut_error("Power up - regulated power reached spec voltage ",
                    "of 650 mV +",pwr_v_tolerance,
                    " before minimum time of 100ns - time measured ",
                    sys.time - pwr_rise_time);
        };
    };
};
```

2.8.4 Comparing a Value to a Threshold

Checking that some signal property is within a range of specified values is another common pattern. Consider, for instance, a voltage regulator needing to maintain its output between minimum and maximum voltage thresholds. In this case, the checking is focused on the dynamic properties of the signal, rather than some statistical average over a period of time.

Checking for threshold crossing is simplified by the *dms_threshold* monitor block. When using the threshold monitor you need to provide the threshold value along with the tolerance. If desired, a suppression time window can be provided, so that jitter lasting less than the specified time window is suppressed (not reported).

Figure 2-16 Threshold Crossing Checker

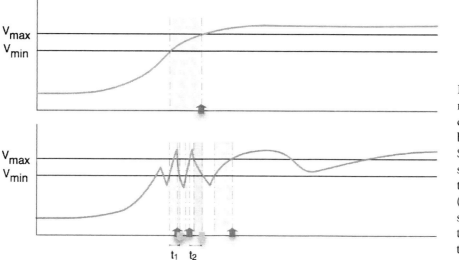

For spikes, multiple crossings may be reported. Spikes are suppressed if their duration (t1, t2) is smaller than the suppression time specifies.

For a minimum and maximum threshold values, as in the voltage regulator example, two threshold detector blocks should be used, one for each threshold. Each monitor should trigger a separate check, such that the error message can be specific and the checks can be controlled (turned off, for example) independently.

It is often the case that the threshold requirement needs to hold within a certain time window after a global event, like reset or power-up. In this case the checker needs to be turned on after the specified delay, otherwise spurious errors will result. This can be achieved by adding an enable event that is triggered and delayed as specified. In order to support power shut-off or reset during simulation, a similar mechanism is needed to turn the checker off.

Example 2–13 *e* Code Using Threshold Crossing Monitors to Check Voltage-Level Range, Ignoring an Initial Power-Up Period

```
unit tb_vreg_checker_u {
    p_smp        :tb_smp_u;  -- reference to signal map
    delta_v      :real;      -- threshold crossing tolerance
    keep delta_v == 50.0;    -- mV

    comp_max :dms_thold_env_u is instance; -- threshold monitor 1
    keep comp_max.hdl_path() == "~/tb_top/thold1";
    comp_min :dms_thold_env_u is instance; -- threshold monitor 2
    keep comp_min.hdl_path() == "~/tb_top/thold2";
    -- derive local events for relevant crossings
    min_crossed_ev   :in event_port is instance;
    keep bind(min_crossed_ev,comp_min.thold_cross_on_fall);
    max_crossed_ev   :in event_port is instance;
    keep bind(max_crossed_ev,comp_max.thold_cross_on_rise);
    -- compute coast signal to ignore power-up period
    !coast   :bool;
    event power_up_start is fall(p_smp.power_down$)@sim;
    event power_up_end is {@power_up_start; delay(100)};
    on power_up_start{
        coast = TRUE;
    };
    on power_up_start_end{
        coast = FALSE;
    };

    configure_checker(min_v :real; max_v :real) is {
        -- called by TB to configure checker
        -- both minimum and maximum values are needed for each
        -- comparator to ensure a crossing has occurred
        comp_min.thold_min = min_v - delta_v;
        comp_min.thold_max = min_v + delta_v;
        comp_max.thold_min = max_v - delta_v;
        comp_max.thold_max = max_v + delta_v;
    };

    event min_crossed is @min_crossed_ev$;
    on min_crossed {
        check that coast else
        dut_error("Power regulator fell below threshold ",
                comp_min.thold_min);
    };
    event max_crossed is @max_crossed_ev$;
    on max_crossed {
        check that coast else
        dut_error("Power regulator rose above threshold ",
                comp_max.thold_max);
    };
};
```

2.8.5 Checking Frequency Response

Some types of analysis are not directly available when running a transient simulation, like the simulation performed by the Cadence Incisive Enterprise Simulator (IES). A typical example is the frequency response of a circuit. While it is not possible to provide an accurate analysis like the one performed by the Cadence Analog Design Environment (ADE), it is possible to approximate the frequency response by sweeping the frequency used as an input to the circuit.

It may be practical to run a number of frequencies in a single simulation, setting a frequency, measuring the response, and switching to the next frequency. Alternatively the frequency can be set for a particular simulation and a higher-level control, like the Cadence Incisive Enterprise Manager (IEM), can be used to set a different frequency to each simulation. In either case, each simulation should be self-checking. The built in checkers should receive the target frequency and measure the response, for example the output amplitude and comparing with the expected result for the given frequency. Expected results should be coded into the checker as a table or a function. See "Creating a Predictor" on page 48 for details.

Because the frequencies are applied one at a time, selecting a good set of test frequencies is important. Key considerations are design spec noted values, like the expected input signal band. Additional frequencies should target specific filter corners (just below and just above the corner frequency). Randomization may be used to provide a wider spread of frequencies and ensure robustness.

Example 2–14 *e* **Code Featuring Top-Level Sequence that Applies a Band of Frequencies Around a Central Operating Frequency**

This portion deals with driving the frequencies, checking code is not included here.

```
extend FREQ_SETUP dms_tb_sequence {
    -- Sequence to program signal generator for specific frequency
    frequency :real;
        keep soft frequency == 1e8;

    body() @driver.clock is only {
        message(HIGH, "Signal source set up, frequency = ", dec(frequency));
        do DRIVE_SEQ dms_wire_seq on driver.sig_in keeping {
            .ampl == 0.005;      -- 1 mV
            .bias == 1.1;
            .freq == frequency;
            .phase == 0.0;
        };
    };
};

extend MAIN dms_tb_sequence {
    -- top-level sequence driving the test
    freq  :real;
    keep freq == rdist_uniform(0.9e9, 2.4e9); -- pick a center frequency
    body() @driver.clock is only {
        wait [2];
        do FREQ_SETUP dms_tb_seq keeping { .frequency == freq; };
```

```
            wait [2];
            out("Power up");
            do PWR_UP dms_pu_trans on driver.power_drv;
            wait fall(driver.smp.power_down$)@sim;
            out("Power-up done");
            for each (f1) in { freq*0.5; freq *0.9; freq; freq*1.1; freq*1.5 } {
                -- set up signal source
                do FREQ_SETUP dms_tb_seq keeping { .frequency == f1; };
                wait [50];
            };
            out("Powering down");
            do PWR_DN dms_pu_trans on driver.power_drv;
            wait [10];
            out("Done");
            stop_run();
        };
    };
```

2.8.5.1 Creating a Predictor

Complex checking often requires a golden reference for comparison. In the case of analog functionality, such a reference may be an equation or a table. When digital controls affect analog properties, a lookup table can be used to predict the functional behavior under given control settings.

Example 2–15 *e* **Code Example Implementing a Predictor**

The predictor is reading the control settings and computes an expected gain based on equation found in the spec.

```
#define GAIN_DB_PER_LSB 0.053
unit vga_checker_u {
    p_env   :dms_tb_env_u;   -- reference to top-level env
    predict_gain():real is {
        -- This predictor method looks up the current control
        -- values and computes the expected gain
        var vga_lsb :int (bits :3) =
            env.vga_control.get_val("vga_lsb"); -- read API for control reg
        var expected_gain_lsb :real;
        var expected_gain_db  :real;
        var expected_gain     :real;
        -- lookup gain factor in table
        case (vga_lsb) {
            3'b000: {expected_gain_lsb = 0.0};
            3'b001: {expected_gain_lsb = 1.1};
            3'b010: {expected_gain_lsb = 2.1};
            3'b011: {expected_gain_lsb = 3.2};
            3'b100: {expected_gain_lsb = 4.2};
            3'b101: {expected_gain_lsb = 5.1};
            3'b110: {expected_gain_lsb = 6.3};
```

```
            3'b111: {expected_gain_lsb = 7.5};
            default: {expected_gain_lsb = 0.0};
        };
        -- compute gain formula
        expected_gain_db = GAIN_DB_PER_LSB * expected_gain_lsb;
        expected_gain = pow(10.0, expected_gain_db/20);
        return expected_gain;
    };
};
```

2.9 Using Assertions

Assertions are a complementary approach to checking. Assertions tend to be local checks that are embedded or associated with the design IP. Assertions are seamlessly integrated into the coverage and reporting scheme of the testbench. Assertions are most often used to check **input conditions** and **local invariants**.

2.9.1 Checking Input Conditions

A common problem encountered during SoC integration is hooking up the IPs incorrectly. This may be as crude as switching the polarity of a signal, or a subtle mismatch in voltage levels. However, once such an error occurs it is very hard to detect and debug in the context of the integrated system. Input checking is all about detecting such problems. They are typically implemented as assertions so that they are included in the integrated system.

Assertion languages like the Property Specification Language (PSL) and SystemVerilog Assertions (SVA) are synchronous, meaning they are associated with a clock event. Still, to check that a condition holds at the end of reset, the falling edge of reset can be used as the clock event. The Cadence Analog Mixed-Signal (AMS) extension to PSL supports checking electrical properties of signals, like the voltage level of a node. Hence verifying the input voltage on a node directly after reset is trivially simple, as shown below.

Example 2-16 PSL Formula for Checking Voltage Levels at Specific Point in Time

```
input_v_check: assert always ((V(sig_in)>0.5m) && (V(sig_in)<10m)) @(negedge reset);
```

A more complicated case arises when the property to be checked is dependent on the configuration. Assume for example, an input is allowed to be 0 if a particular feature is disabled, but should meet the specified voltage range otherwise. This can be expressed using an implication (an overlapping suffix implication in PSL).

Example 2-17 PSL Formula for Conditional Checking at Specific Point in Time

```
input_v_cond_check: assert always feature_enabled |->((V(sig_in)>0.5m) && (V(sig_in)<10m)) @(negedge reset);
```

Note Don't use |=> (PSL non-overlapping suffix implication) if the clock expression is aperiodic, like reset. This means the condition will be checked the next time reset is deasserted!

2.9.2 Verifying Local Invariants

Assertions within the core of the IP block may help expose subtle errors. They often express the designer expectation about a certain condition always holding true. Such conditions are called invariants. For example, a circuit may be designed to blank out if the input voltage is greater than a threshold. One may assert that the blank out signal is high if the voltage is above the specified threshold. Note that this only works for slow changing signals (otherwise there is no guarantee the signal is sampled at the peak). The following PSL formula uses the default clock to time the checking.

Example 2–18 PSL Formula that Checks for an Invariant Condition

```
default clock = (posedge clk);
blankout_chk: assert always (V(input_sig)>threshold)|-> blank_out;
```

2.9.3 Limitations on Assertion Checking of Analog Properties

The example above has a discrete event (the clock) as the trigger condition. Checking the property for the un-sampled input signal requires continuous monitoring of the input signal voltage, which is not very natural for assertion languages. As mentioned both PSL and SVA require a sampling clock. In essence, all checks use sampled values of the analog signals, and there may not be a suitable sampling clock available. Even if such a clock exists, checking a condition continuously may have a significant performance impact. Such cases are better handled by testbench-level checking as discussed in section 4.5.

2.9.4 Dealing with Different Modeling Styles

Assertions sample analog values by using the modeling language capabilities, hence their syntax may need to change when switching between a netlist model, Verilog-AMS, and real-number models. Since assertions are often embedded within the model file, they would be duplicated naturally. In other cases, the assertion must be implemented in several styles under a conditional macro. Creating a clear separation between the asserted property and the hookup is possible using the *vuint* construct. Consider the code in the following example. Both module types (the one called *intf_analog* and the one called *intf_ams*) contain the signals *power_down* and *reset* in their scope.

Example 2–19 PSL Verification Unit that is Bound to diFferent Modules Depending on the Model Used

```
vuint input_monitor (
`ifdef USE_SPICE_NETLIST
    intf_analog
`else
    intf_ams
`endif
) {
  property positive_power_down_at_reset =
    (V(power_down) > 0) @(negedge reset);
  a1: assert positive_power_down_at_reset;
}
```

Note PSL verification units can only connect to module types at present. Connection to specific instances is not supported. Also note that coverage of a PSL assertion embedded within an AMS block is only supported in IES 10.2.

2.10 Clocks, Resets and Power Controls

Analog IP blocks may contain some digital controls that require clock and reset signals to be driven by the testbench. Even in cases where the design is purely analog, the testbench around the circuit will probably require a clock and a reset. To ease the integration into an SoC environment, the test environment should include a clear interface for clocks and resets.

2.10.1 Driving Clocks

The test environment should be modeled after the actual integration environment within the SoC. If the SoC provides several clocks and reset signals that may be relevant, it is best to represent them all. Clocks may be gated, for instance, and the testbench should make it possible to replicate the same conditions at the IP level.

In addition to the clocks driving the design, there will be clocks used in the testbench, for example to sample analog signals. It is important to keep these independent. Design clocks will eventually be tied to the SoC clock signals. Testbench clocks need to run fast enough to provide adequate sampling of the signals being monitored. Confusing the two functions may prevent checkers and coverage monitors from functioning correctly. Determining and driving testbench clocks is discussed further in "Integrating the Test Environment" on page 54.

2.10.2 Resets

Reset signals are used to initialize the design. The same signals may be used to initialize testbench registers. Another important role of reset is to prevent testbench errors from being reported while the design is being initialized.

It is common to have a hard-wired reset period at a fixed time during the start of simulation. A better implementation will allow the test writer explicit control over reset, through a sequence item for example. Tests that apply reset at some point during simulation may uncover issues with recovery that would otherwise be ignored.

2.10.3 Power-Up and Power-Down Sequences

It is not uncommon to have several supply voltages provided to an IP block, possibly coming from different power domains. Powering up such a circuit may require a specific order and timing. This is called the power-up sequence. Similarly, the circuit may be powered down completely, or put into sleep mode by switching some power signals off in a specific order. This brings about three kinds of requirements for the testbench:

- The testbench must be able to drive the required sequences.
- The testbench must check that the sequences are applied as specified.

- The testbench must collect coverage from the specified sequences.

Power sequences that are not entirely trivial should be driven using a UVM architecture including a BFM, a sequence driver, and a library of sequence items. This will provide the greatest flexibility to the test writer to combine power sequences along with other test sequences. An example of a simple power sequence controller is provided in Figure 2-9 and Example 2-6 on page 34.

Checking that the power-up/down sequence applied to the circuit meets requirements can be done using assertions if the relationship is simple enough to express in PSL. In case a sequence of specific timing and voltage values needs to be checked, it might be more practical to create a dedicated checker in the test environment using a verification language.

Monitoring power-up sequence coverage can leverage events defined by the power sequence checker. Counting the occurrences of various power sequences should be sufficient. If there are some major modes that interact with power behavior (for example a "no-sleep" mode), cross coverage between the mode and the power sequence may make sense.

Figure 2-17   A Simple Power-Up Spec

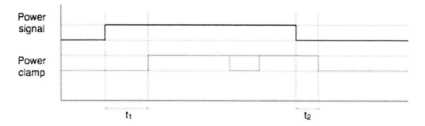

In this figure, power clamp should assert after power signal is received, and must not de-assert until power signal is lost. The response times t1 and t2 are subject to coverage.

Example 2–20   e Code Example Combining Checking and Coverage of a Simple Power Sequence

```
// -------------------------------------------------
    // power-up / down checker
    // -------------------------------------------------
    p_smp   :tb_smp_u;  --reference to signal map
    !pwr_clamp_rise_time    :time;
    !pwr_signal_rise_time   :time;
    !pwr_clamp_fall_time    :time;
    !pwr_signal_fall_time   :time;
    event pwr_signal_rise is rise(p_smp.powerup$)@sim;
    on pwr_signal_rise {
        pwr_signal_rise_time = sys.time;
    };

    event pwr_clamp_rise is rise(p_smp.power_clamp$)@sim;
    on pwr_clamp_rise {
```

```
            pwr_clamp_rise_time = sys.time;
            check that pwr_signal_rise_time > 0 and
                       pwr_signal_rise_time < pwr_clamp_rise_time else
            dut_error("power-up sequence - Power clamp rise detected before power
  signal");
        };

        event pwr_signal_fall is fall(p_smp.powerup$)@sim;
        on pwr_signal_fall {
            pwr_signal_fall_time = sys.time;
        };

        event pwr_clamp_fall is fall(p_smp.power_clamp$)@sim;
        on pwr_clamp_fall {
            pwr_clamp_fall_time = sys.time;
            check that pwr_signal_fall_time > 0 and
                       pwr_signal_fall_time > pwr_signal_rise_time else
            dut_error("power-up sequence - Power clamp fall while power signal is
  high");
        };

        // -------------------------------------------------
        // power-up / down coverage
        // -------------------------------------------------
        cover pwr_clamp_rise is {
            -- compute the delay between power signal rise and power clamp rise
            item clamp_gain_time : int =
                pwr_clamp_rise_time - pwr_signal_rise_time;
        };
        cover pwr_clamp_fall is {
            -- compute the delay between power signal fall and power clamp rise
            item clamp_loss_time : int =
                pwr_clamp_fall_time - pwr_signal_fall_time using
                    ignore = (signal_to_clamp_loss_time < 0);
        };
    };
```

2.11 Analog Model Creation and Validation

Analog models are often developed at a concrete netlist level, using Cadence Spectre (or some other variant of SPICE) to simulate. These models are highly accurate, but they are very slow to simulate. This is a major drawback when considering the large number of combinations that have to be covered to verify functionality.

Behavioral analog models, like Verilog-AMS offer a much faster runtime, while compromising accuracy to some small degree. Much analog analysis can still be performed on AMS models, as they are simulated using an analog solver.

Even better performance can be achieved using Real Number Models—Verilog models using *reals* and *wreals*[1] for example (these are called these *wreal* models here). Such models are even more abstract, as the user chooses explicitly which electrical phenomenons to model. Such a modeling style can be executed on an event-driven simulator, which is orders of magnitude faster than using an analog solver. Such models tend to be less accurate and they can only be used for functional simulation (transient analysis).

In many cases a wrcal model is the best choice for implementing the UVM-MS. Model accuracy is typically not an issue for the kind of checking done during functional simulations and the speed advantage is very significant. Creating such a model may require a considerable investment, but that investment will be leveraged for IP verification and possibly multiple SoC integrations (where speed is really critical). Furthermore, such models can be reused with small modifications from one generation of a design to the next.

The UVM-MS methodology applies to all modeling styles, enabling easy switching of models within the same test environment. This is especially important for debug purposes, when an error detected during a fast model simulation needs to be replayed on a more accurate model.

2.12 Integrating the Test Environment

Test environment integration includes hooking up the testbench with the design model and creating the necessary run scripts to perform a simulation run. This chapter discusses the specifics of carrying out the integration.

The integrated environment should meet the following requirements:
- Simple invocation of a simulation run, in either batch or interactive modes.
- Easy switching of test scenarios.
- Easy switching of design models (from wreal, to AMS, to Spice netlist).
- Ready for integration with EnterpriseManager.
- Reusable at the SoC level and for future projects.

2.12.1 Connecting the Testbench

The UVM requires that all connectivity between the testbench and the design be done through a signal map. A signal map is a unit of the testbench that contains all the external interfaces. The benefit of keeping all connections in a single unit is reusability—when the design changes, some connection paths may change. With a signal map architecture all these changes are localized in one unit, so it is easy to switch back and forth between configurations by maintaining several versions of the signal map.

Analog interfaces have specific hookup requirements, some of which are language dependent. In *e*, all connections to the analog domain must be done through ports.

---

1. Verilog real value wires: Cadence extended the definition of wreals to make a practical real-number modeling approach using Verilog (digital only).

2.12.2 Connecting to Electrical Nodes

A port of type *real* can connect to an electrical node. An *HDL path* attribute must be provided, specifying the node location. The analog quantity read or written must be specified using the *analog kind* attribute.

Example 2–21 *e* **Code Example, Connecting to Electrical Nodes**

```
unit tb_smp_u {
    ref_vol        :inout simple_port of real is instance;
    keep ref_vol.hdl_path()=="~/tb_top/ref";
    keep ref_vol.analog_kind() == potential;
    ref_cur        :inout simple_port of real is instance;
    keep ref_cur.hdl_path()=="~/tb_top/ref";
    keep ref_vol.analog_kind() == flow;
};
```

2.12.2.1 Connecting to Real-Value Nodes

Connections can be made to ports of type *real*. An *HDL path* attribute must be provided, specifying the node location. If the node is a wire, a *wire* attribute has to be set.

Example 2–22 *e* **Code Example of Connections to Real Register and Real Wire (*wreal*)**

```
unit tb_smp_u {
    real_vga_gain_db : in simple_port of real is instance;
    keep bind(real_vga_gain_db, external);
    keep real_vga_gain_db.hdl_path() == "~/tb_top/raw_gain_db";
    real_vga_bias : in simple_port of real is instance;
    keep bind(real_vga_bias, external);
    -- real_vga_bias is connected to a wreal wire
    keep real_vga_bias.hdl_path() == "~/tb_top/raw_bias";
    keep real_vga_bias.verilog_wire() == TRUE;
};
```

2.12.2.2 Connecting to Single-Bit (Logic) Nodes

The testbench can connect to ports of type *bit*. A *wire* attribute must be used if the node is a wire. This is the same as connecting to nodes in the digital portion of the design.

2.12.2.3 Connecting to Expressions

A port can connect to an expression rather than a node in the analog domain. The expression is interpreted in the context of the analog domain (so it needs to be a legal syntax in that domain language). The connecting port may be a real value port, a bit (**B**oolean) or an event port—all depending on the type returned by the expression. The examples below connects to a Verilog-AMS node.

Example 2–23 *e* **Code Example of a Port Connected to an Expression in Verilog-AMS Domain**

```
unit tb_smp_u {
    voltage_diff : in simple_port of real is instance;
    keep voltage_diff.hdl_expression() == "V(tb_top.vin_p)-V(tb_top.vin_n)";
    keep voltage_diff.analog() == TRUE;
};
```

2.12.3 System-Level Parameters and Timing

The testbench and the various design models are typically developed independently, possibly by different people. They may be using different values for properties such as supply voltages, clock frequencies, and other model parameters. An important part of integrating the environment is creating a consistent top-level set of parameters that serve as a central point of control. Most of these parameters are available in a simulation control file for the netlist and AMS models. Some of these parameters need to be set by the testbench. The testbench may need access to some of the simulation control parameters, in which case it should read those upon start-up.

Parameter setting should be made easily visible and controllable. An example is provided in Example 2–5 on page 32. As far as visibility is concerned, collecting configuration parameters such as coverage, and reporting them to log files are recommended practices.

Though the design model may be continuous in time, the higher levels of the testbench are always executed in a discrete, event-driven manner. This means analog signals are sampled and some inputs are generated using discrete time steps. To ensure good accuracy of operation, it is critical that driving and sampling are synchronized across the entire testbench. In case of real-number models that are also event driven, the model clocking has to be similarly synchronized. This issue is most pressing at the SoC level, where complex timing behavior of the whole system needs to be resolved.

Figure 2-18 The Effects of Sampling a Quantized Signal

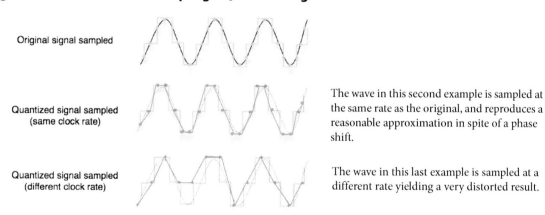

Timing synchronization can be achieved by generating a single fastest clock at the testbench level and deriving any local clocks from it. If the testbench or the design includes some self-contained units that generate their own clock, a common control parameter should determine the clock frequency, and the activation of the units has to be synchronized such that the overall timing behavior is globally synchronous.

Because testing requirements may involve dynamically varying frequencies and modulations, the selection of the fastest clock frequency needs to be flexible. As a general rule, you should allow at least 10 samples per cycle for the fastest signal, else sampling and generation will yield very noisy signals. If the design includes any filters modeled in the time domain, such filters will not function correctly if under sampled. On the other hand, over sampling at a very high rate will slow simulation down without any notable benefit in accuracy. It is best that the sampling frequency be set at the beginning of simulation, to match the range of frequencies that are being used in a specific test.

2.12.4 Supporting Several Model Styles In A Single Testbench

An important feature of any analog testbench is the ability to switch between netlist, AMS, and real-number models. Such a testbench can be used to validate one model against the other by running the same simulation on both. This is also critically important for debugging, when a low accuracy model reports an error, which has to be reproduced and debugged using a high accuracy model.

The simulation environment offers some support for switching models in a seamless way. Additional methodological guidance is provided in this section to simplify the task.

2.12.4.1 Switching Models Using Define Macros

The verification environment should use a global set of define macros as the primary mechanism for switching models. Setting the define macro at the run script would be the mechanism for switching models.

The macro switch is used to switch the type of the DUT model instantiated. It is also used to change the definition of interface signals that are impacted by the change. Various passive elements that may be required for high accuracy models are left out for real-number models. Those can also be switched in and out using the macro. The following code listing is an example of such a switchable model.

Example 2-24 Verilog Code—Netlist Elements and Real-Number Modeling

Verilog code containing both netlist elements and real-number modeling. The active parts are switchable by a define macro.

```
    module dms_tb_top;
      // ---- Declarations --------------------
    `ifdef USE_ELECTRICAL_MODEL
      // everything needed for AMS / Spice sim
      electrical osc1_p, osc1_n, noise, wr;
      electrical ain_p, ain_n;
      electrical aout;
      electrical gnda;
      ground gnda;
    `else
```

```
    // the abstract WREAL model
    wreal osc1_p, osc1_n, noise, wr;
    wreal ain_p, ain_n;
    wreal aout;
`endif

    // ---- Combining noise with signal -----------
`ifdef USE_ELECTRICAL_MODEL
    // Passive and active elements
    resistor #(.r(10M)) R1 (noise, gnda);
    vcvs #(.gain(1.0)) IS_1 (ain_p, osc1_p, noise, gnda );
    vcvs #(.gain(1.0)) IS_2 (ain_n, osc1_n, noise, gnda );
`else
    // Equivalent function
    assign ain_p = osc1_p + noise;
    assign ain_n = osc1_n + noise;
`endif

    // ---- Analog DUT ------------------
    ncamp_dut ncamp(.ain_p(ain_p), .ain_n(ain_n), .aout(aout) );
endmodule
```

Though in the simple example above there is little sharing between the two modes of the testbench, in more realistic examples a much larger portion is shared between the modes. This is more efficient and easier to maintain than two independent HDL testbenches, one for each model type.

The higher levels of the testbench may also require minor changes. Some connections in the signal map may need to change based on the target model. Some checks and coverage points may need to be deactivated because nodes internal to the circuit cannot be accessed in a netlist model. These restrictions were discussed in previous sections.

While using `define is possible within high-level verification language (HVL) testbenches, it is much better to use dynamic subtypes (also called *when inheritance*) when using *e*. With dynamic sub-types, the compiler accesses the whole of the code and can identify errors that would be otherwise hidden. The macro is used once at the top level to set the environment sub-type.

Example 2–25 *e* **Code Example—Dynamic Sub-Type of Testbench**

e code example for creating a dynamic sub-type of the testbench code and propagating the mode in effect from a command line define through the testbench hierarchy to the point of use within the signal map.

```
    #ifdef USE_ELECTRICAL_MODEL then {
        extend sys {
            keep tb_env.kind == INTF_ONLY_VISIBILITY;
        };
    };

    -- The following type defines the dynamic sub-types
    type tb_kind_t: [INTF_ONLY_VISIBILITY, FULL_VISIBILITY];
```

```
unit tb_env_u {
    kind :tb_kind_t;
    keep soft kind == FULL_VISIBILITY;
    smp  :tb_smp_u is instance;
    keep smp.kind == kind; -- propagate kind down the hierarchy
};

unit tb_smp_u {
    kind    :tb_kind_t;
    cmp_t   :in simple_port of real is instance;
    keep bind((cmp_t, external);
    -- For Spice netlist - there is a E2E connection
    -- so the following lines would be needed
    when INTF_ONLY_VISIBILITY tb_smp_u {
        keep cmp_t.hdl_path()== "~/tb_top/cmp_t_reg";
    };
    when FULL_VISIBILITY tb_smp_u {
        keep cmp_t.hdl_path()== "~/tb_top/cmp_t";
        keep cmp_t.verilog_wire()== TRUE;
    };
};
```

Example 2-26   Command-Line Options for irun

Command line options specified for irun to choose a SPICE netlist model as defined in the previous examples:

```
-define   USE_ELECTRICAL_MODEL
-sndefine USE_ELECTRICAL_MODEL
```

2.12.5   Interfacing Between Real and Electrical Signals

The testbench most often uses discrete domain values to represent signals, which could be real or Boolean values. When the design is modeled as an AMS or netlist model, its terminals are in the continuous domain, either *logic* or *electrical* discipline. The elaboration process in the simulator applies a resolution algorithm to determine the default discipline for each network, and if the network is crossing between the continuous and discrete domains, a *connect module* is automatically inserted to bridge the two.

Automatically inserted connect modules may require tuning. By default, the connect module will determine the threshold between logic 1 and logic 0 based on the supply voltage. Similarly, the electrical voltage resolution will be computed based on the supply voltage, typically V/64 (dividing the supply range to 64 voltage increments). This may work in case the signal makes use of the full dynamic range, but when the input signal is very small when compared with the supply voltage, the converted signal may be very noisy or absent altogether.

Figure 2-19 Connect Module Resolution Derived from Supply Voltage

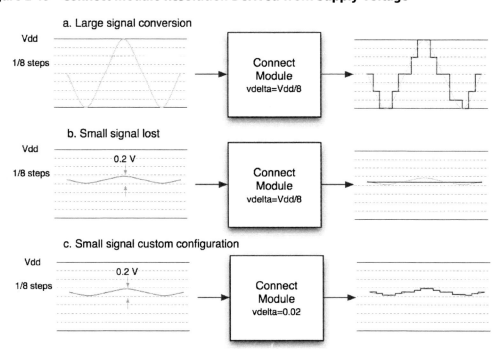

In the above figure, connect module resolution is derived from the supply voltage by default. A large signal (a) is converted naturally, but a small signal (b) may be lost altogether. Tuning the connect module *vdelta* parameter (c) recovers the small signal.

Connect modules can be tuned globally by setting their parameters in the *interface element* (*ie*) statement of the *amsd* control block of the *simulator control file*. For Spectre, this file is called *amscf_spice.scs*, and it should be included in the *irun* source list along other HDL sources. An example is provided below.

Example 2-27 Using the Simulation Control File *amscf_spice.scs* to Control Connect Module Features

```
simulator lang=spectre
//<... Content Omitted...>

amsd {
    // For real values, controlling voltage resolution
    ie vsup=1.0 vdelta=1u // supply voltage 1V, resolution 1uV
    // For logic signals
    ie vsup=1.0 vthi=0.4 vtlo=0.2 // Specify high and low thresholds
    // control a specific scope
    ie vsup=1.8 inst="top.driver"
}
```

2.12.6 Creating Run Scripts and Other Support Files

Cadence suggests that a simulation run file be created as the way to launch a simulation. Different run files should be created to run with different models, or to run in debug versus batch mode. Since all these files are simple shell scripts, you may create a common interface or combine them into a single script controlled by command line options as desired. A simple run file using the IES `irun` command is shown below. The example specifies a *run.tcl* file. That file should include environment setup and configuration options that do not affect model construction.

Example 2-28   A Simple run_sim File Invoking irun

Parameters to the script are passed directly to irun.

```
#!/bin/csh -f
# File: Run_sim
irun -f run.f -gui -input run.tcl \
     -timescale 1ns/100ps -access +rwc $1
```

The Cadence SimVision debugging tool can create a configuration file (called *restore.tcl* by default) to configure the IES runtime environment. That file is specified in the example below, to invoke the debug environment as it was saved last.

Example 2-29   Minimal run.tcl File Collecting Waveforms During Run

```
#File: run.tcl
# collect waveforms
probe -create -shm dms_top -depth all -all
# invoke latest IES configuration
input restore.tcl
```

The following example *Run_sim* script calls a file called *run.f* which is used to control the model build options.

Example 2-30   A Sample run.f File Used to Define the Build and Run Process

```
# File: run.f
# Specman settings
# Must use IntelliGen in order to generate real types
-snset "config gen   -default_generator=IntelliGen"
-snset "config cover -write_model=ucm"  # configure coverage
-snset "config cover -database_format=ucd"
-snset "config run -tick_max=1M"# Specman runtime limit
-sntimescale "1ns/1ns"# Specman time scale
-access rwc# testbench access rights
-wreal_resolution sum# WREAL resolution function

-define   USE_ELECTRICAL_MODEL# control which model executes
# Sources to include in build
$WORKAREA/hdl/dms/dms_mswire_uvc.vams
$WORKAREA/hdl/tb/dms_noisy_src.vams
```

```
$WORKAREA/hdl/tb/amscf_spice.scs

-nosncomp     # do not compile e files (optional)
$WORKAREA/e/dms_tb_top.e    # TB top-level file
# the following invokes a local command file that loads the selected test
# the name is important - this is how EManager integrates and controls the
# run
-snprerun "@local_ecom.ecom"
```

2.12.7 Recommended Directory Structure

The use of a standard directory structure is recommended, as it helps orientation in unfamiliar projects. The proposed structure maintains separation between design and testbench code, as well as test planning and documentation. Such modularity is important to facilitate reuse and integration. The following is an example listing of a standard project directory.

Table 2-1 Recommended Structure for Top-Level Project Directory

Directory or File	Contents
`docs`	Documentation, such as specs, design documents
`e`	Testbench *e* code
`hdl/`	Design code
`assertion`	Assertions code
`cm`	Customized connect modules
`spice`	Spice netlist model
`vams`	Verilog-AMS model
`wreal`	Verilog real-number model
`uvm_dms_lib`	Library of reusable verification components
`sim`	Simulation run directory, regression runs.
`tb`	HDL testbench code
`vplan/`	Verification planning, Enterprise Planner files.
`analysis`	Enterprise Planner files
`execute`	vsif files
`setup.csh`	A shell script setting up the work environment

The various compile and run scripts mentioned in the previous section are located under the *sim* directory. Low-level testbench code implemented in HDL (Verilog AMS for example) is located in the *tb* directory.

2.13 Closing the Loop Between Regressions and Plan

The section "Including Analog Properties" on page 20 describes the creation of an executable verification plan (vPlan) using Enterprise Planner for analog properties of the DUT. The high-level structure of the plan is created first, and all features of interest are categorized into appropriate sub-sections. Appropriate sections are also created for planned coverage collection. Similarly, planned checks and scenarios are also added to the vPlan. This forms the basis for detailed implementation of the verification environment. During the implementation stage cover groups are created in the testbench. Coverage will be captured and mapped as specified in the vPlan. There is an implicit correlation between this raw collected coverage data and the planned coverage sub-sections in the vPlan. However, there is no direct link between the two. This task needs to be performed manually, whereby the loop is closed between the planned coverage in the vPlan and the raw coverage data obtained from regression runs. This process is described in the following section. Updating the verification plan with implementation measured coverage provides the most useful feedback about the state of the verification effort.

2.13.1 Implementation of Coverage for Analog

Example 2–31 on page 63 shows an example of cover groups implemented in the verification environment (VE). This code generates functional coverage data for analog as shown in Figure 2-20 on page 64.

Example 2–31 *e* **Code for Cover Groups Implemented in Testbench**

```
unit gain_mon_u {
  event gain_sample is @smp.rd_clk_fall;
  cover gain_sample is {
    item real_vga_gain_db : real =
        smp.as_a(FULL_VISIBILITY tb_smp).real_vga_gain_db$ using ranges = {
            range([22.5..25.5],   "Real: VGA-GAIN = 24dB +/- 1.5dB" );
            range([19.5..22.5],   "Real: VGA-GAIN = 21dB +/- 1.5dB" );
            range([16.5..19.5],   "Real: VGA-GAIN = 18dB +/- 1.5dB" );
            range([13.5..16.5],   "Real: VGA-GAIN = 15dB +/- 1.5dB" );
            range([10.5..13.5],   "Real: VGA-GAIN = 12dB +/- 1.5dB" );
            range([7.5..10.5],    "Real: VGA-GAIN =  9dB +/- 1.5dB" );
            range([4.5..7.5],     "Real: VGA-GAIN =  6dB +/- 1.5dB" );
            range([1.5..4.5],     "Real: VGA-GAIN =  3dB +/- 1.5dB" );
            range([0.0..1.5],     "Real: VGA-GAIN =  0dB +/- 1.5dB" );
        }, illegal = (real_vga_gain_db > 26.5);

    item sig_vga_control1 : uint(bits:4) =
        smp.sig_vga_control_reg1$ using ranges =   {
            range ([4'b0001],   "Sig: sig_vga_control1 = 0" );
            range ([4'b0011],   "Sig: sig_vga_control1 = 8" );
            range ([4'b0111],   "Sig: sig_vga_control1 = 16" );
            range ([4'b1111],   "Sig: sig_vga_control1 = 24" );
        };
  }; // cover
}; // unit
```

Figure 2-20 Unmapped Analog Coverage Data from Regression Runs

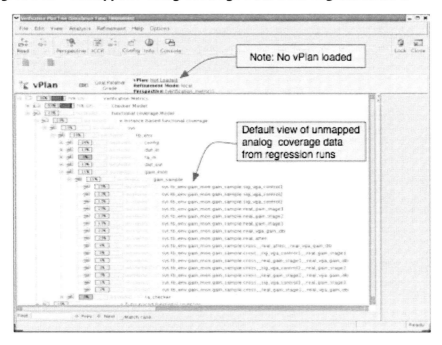

The figure shows coverage data collected by the VE—this is the default view in Incisive Enterprise Manager. Note that there is no vPlan loaded, hence this is an unmapped view of the collected metrics. The results are split into separate categories representing metrics from test scenarios, checkers, and functional and code coverage. The functional coverage leaf-data represents the coverage items directly from cover groups in the VE. These metrics will be mapped into the vPlan using Enterprise Planner as shown in the next section.

2.13.2 Updating the Verification Plan With Implementation Data

The initial vPlan, shown in Figure 2-21 on page 65, contains sections for analog coverage. These are place holders for planned coverage which initially do not have any collected coverage data. As regressions are run, a lot of raw coverage data is collected and available for analysis, as shown in Figure 2-20 on page 64. This large collected amount of unmapped coverage data are hard to analyze and use meaningfully. The solution is to project the raw coverage data onto the vPlan. This enables coverage data to be gauged in a meaningful and relevant way, as intended during the planning process.

Figure 2-21 Initial vPlan for Analog Components

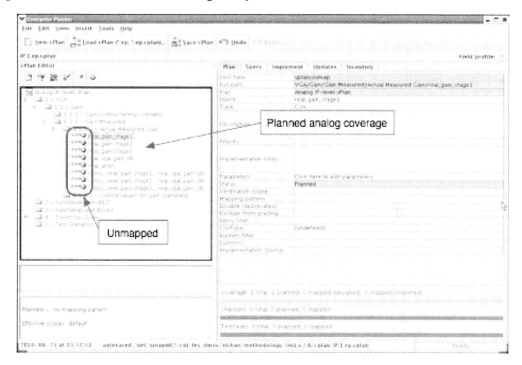

2.14 Regression Runs for Analog IP

2.14.1 Single Simulation Runs

To invoke a single simulation run, use the commands listed in Example 2–32.

Example 2–32 Commands for Invoking a Simulation Run

```
cd <WORKAREA>/sim
./Run_sim_ams
./Run_sim_wreals
./Run_sim_spice
```

Each of the above commands runs a different simulation configuration.

The following two examples present a sample run script hierarchy. Example 2–33 is the main run script, invoking the *irun* batch mode simulation interface. This script is importing a command file provided in Example 2–34 on page 66.

Example 2–33 A Simulation Run File Example

```
#! /bin/csh -f
# File: Run_sim_wreals
irun \
     -input $WORKAREA/sim/run.tcl \
     -f $WORKAREA/sim/run.wreals.f \
     -f $WORKAREA/sim/common_defines.f \
     -incdir $WORKAREA/tb \
     -incdir $WORKAREA/hdl/cm \
     -y $WORKAREA/hdl/lib \
     -timescale 1ns/10ps \
     -access +rwc \
     -write_metrics \
     -nclibdirname $BUILD_DIR \
     $1
```

Example 2–34 An Example Build and Run Command File

```
# File: run.wreals.f
## Design & TB
$WORKAREA/hdl/vams/cds_globals.vams
$WORKAREA/hdl/lib/*.v
$WORKAREA/hdl/cm/*.vams
$WORKAREA/hdl/wreal/dut_wreal.vams
$WORKAREA/tb/dut_ana_top_tb.vams
$WORKAREA/tb/amscf.scs

# OVM-MS UVC
$WORKAREA/ovm_dms_lib/dms_mswire/hdl/dms/dms_mswire_uvc.vams

# Specman setup
-snset "config gen    -default_generator=IntelliGen"
-snset "config cover  -write_model=ucm"
-snset "config cover  -database_format=ucd"
-snset "config cover  -collect_checks_expects=TRUE"
-snset "config run    -tick_max=1M"
-snprerun "@local_ecom.ecom"   #Note - necessary for IEM integration!
-nosncomp
$WORKAREA/e/tb_top.e

#####################################
# Use Wreal models of DUT
# Note: Both -define and -sndefine
#       MUST be set to the SAME
#####################################
-define    USE_WREAL_MODELS
-sndefine  USE_WREAL_MODELS
```

2.14.2 Regressions—Running Multiple Test Cases

Regressions are submitted, controlled, and analyzed by Incisive Enterprise Manager (IEM). To invoke a regression run, first invoke IEM using the following command-line:

```
emanager -p "start_session -vsif <WORKAREA>/ vplan/regress_wreals.vsif
```

This reads the VSIF file and starts a regression run, which can then be analyzed using IEM. Figure 2-22 depicts the IEM interface that supports tracking and control of the executing simulations. This interface also provides basic debug capabilities in case simulations fail or get stuck.

Figure 2-22   Running Regression using Incisive Enterprise Manager

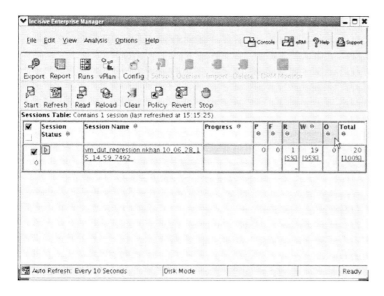

Once the regressions have fully executed, the vPlan can be loaded so that coverage can be analyzed in reference with the original plan. The loading of a vPlan file is illustrated in Figure 2-23 and the resulting mapped coverage is depicted in Figure 2-24.

UVM and Metric-Driven Verification for Mixed-Signal

Figure 2-23   Load vPlan for Analog IP

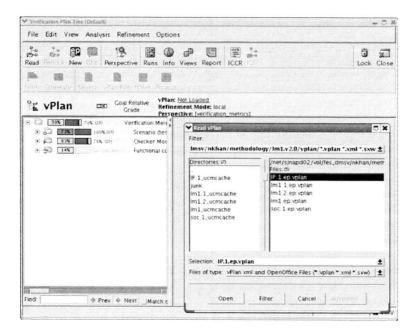

Figure 2-24   Analog Coverage

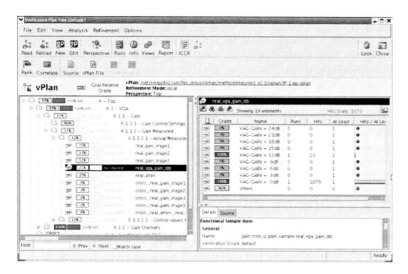

2.15 Moving Up to the SoC Level

Thorough verification at the IP level minimizes the integration risk because the quality of the IP is known. At the same time, the IP-level verification effort can be leveraged at the SoC level, helping to quickly create a sophisticated test environment. This section discusses the benefits of reuse at the SoC level.

Many of the integration issues encountered at the SoC level are similar to the integration of an IP block with its test environment. The issues are mentioned in this section and reference the more detailed discussion presented above.

2.15.1 Mix-and-Match SoC-Level Simulation

Probably the most valuable component to be reused is the model of the IP. Assuming an abstract (AMS or real number) model was developed during IP verification, it can be inserted into the SoC-level simulation, providing a huge speedup in simulation time.

Integrating a netlist model of the IP into the SoC simulation is impractical in many cases—for example each time the SoC includes a processor that has to boot up before the analog interfaces can be activated. Such a mixed-mode simulation can be prohibitively slow, requiring many weeks of simulation time. In these cases, the investment in a good abstract model of the IP is easy to justify.

Even in cases where a Spice netlist simulation can be performed, it should probably only be used only to verify some critical circuit aspects. The bulk of the simulation runs should be done using the abstract model. Models developed according to the methodology presented here should be sufficiently accurate and well verified to be trusted for system-level simulation.

2.15.1.1 Interface Consistency for SoC-Level Model Switching

An SoC can have multiple analog and mixed-signal IP blocks. Various verification goals may require some of these blocks to be represented in full detail, as Spice netlists, while others can be represented as abstract models. Models should have matching interfaces, so that switching models does not require changes to the integrating level (the "socket").

To enable seamless switching of models, all versions of a specific IP must maintain the exact same interface. It is recommended that port names, sizes, and order are kept uniform across all models of a given IP. Even if some interface pins are not used in some case, they should still be declared for compatibility sake. The Spice netlist model is the master model dictating the interface. Avoid "virtual" ports (ports that won't exist on the chip) in the abstract models. Out-of-module references should be used to implement such virtual access points.

The IP-level testbench defines the socket for the IP, matching all IP model interfaces. This topic is covered in some details in "Switching Models Using Define Macros" on page 57. This enables switching of IP models at that level, as well as model validation by running two models of the IP and comparing their outputs. Models developed using this methodology are ensured to have compatible interfaces and are easily switchable at the SoC level.

2.15.1.2 Port Connectivity Coercion When Switching Models

Another aspect to consider is the port type, especially when using real-number models. A port may be defined as a real value (wreal) in an abstract model and as an electrical type in a netlist model. Similarly, a digital interface may be driving a logic value, while the netlist model expects an electrical value.

The Incisive Simulation Platform resolves these interface issues automatically, by inserting the appropriate connect modules to perform a conversion as needed. Any issues would typically be encountered during the IP verification phase, as the IP-level testbench should define a socket similar to the SoC one. Coercion is performed at that level to allow model switching. Tuning of connect modules and typical problems are encountered and addressed, so that the IP is ready for integration. This issue is further discussed in "Interfacing Between Real and Electrical Signals" on page 59.

2.15.2 Updating the SoC-Level Test Plan

The IP test plan can be easily pulled into the SoC-level plan as a chapter in Enterprise Planner (EP). EP has mapping options that support the instantiation of IP, even of multiple instances, though this requires that coverage is collected on an instance-by-instance basis (in contrast to type-based coverage). Please see the Enterprise Planner documentation for a discussion of these features.

Instance-based real number coverage in *e* has limited support: only unit instance-based coverage is supported (but not struct instance coverage).

Plan integration needs to go beyond the mechanical inclusion of the plan. It is unlikely that the complete testing of the IP can be repeated at the SoC level, given the much broader verification needs and the high cost of simulation at the system level. Instead, a careful review of the original plan needs to be carried out. A small subset of core features and items that are likely to be affected by integration should be marked for execution, and the rest of the items need to be disabled. EP provides a disable checkbox on each item, so that they can be excluded in a non-destructive way.

After executing a first regression of the integrated SoC, the verification engineer needs to map the coverage, checks, and scenarios produced by regression back to the integrated vPlan. This process needs to be done with a critical eye, to ensure plan items are not discharged or credited in error.

2.15.3 Integrating Into the SoC-Level Testbench

The benefit of reusing verification components from IP to SoC-level environments is often overlooked. Yet embedded in such components is the most accurate characterization of behavior available, one that was extensively used while developing the IP. Hence, there is a strong case for integrating components of the IP-level testbench.

2.15.3.1 Adding Coverage and Checking

Coverage and checking monitors are easier to integrate, as they are non-intrusive observers. The UVM architecture ensures that integration is simple, because these components are modular and independent. The integration involves the following steps:

1. Include the modules containing the verification code into the SoC-level test environment sources.

2. Instantiate the monitor environment, or the whole UVC, at the top-level SoC environment. If dealing with a UVC, the active behaviors have to be turned off, for instance by constraining the mode to be PASSIVE.

3. Replace the signal map, typically by extending the signal map class and creating a dynamic sub-type. Set the appropriate HDL paths such that the integrated IP module is accessed correctly.

4. Identify any timing sources, such as events that trigger or activate the monitor. Select the appropriate activation condition in the integrated environment. Most often this is resolved by a signal map assignment (such as finding the right signal to monitor for changes). Sometimes, however, this may require some supporting code, such as creating some logic in the testbench to identify the condition required.

Before attempting the integration, however, you have to consider the impact of the monitor on simulation performance, keeping in mind that the SoC environment is much more performance intensive. Consult the verification plan to ensure that the coverage and checking collected by the monitor is considered important enough for the integration. Try to estimate the impact in terms of *number of events per simulation*.

The functionality of monitors created following the UVM can be changed dynamically. Checking and coverage collection can be turned off using constraints (by setting the *has_checks* and *has_coverage* modes to FALSE). Therefore, it is suggested that most checkers and coverage be integrated, and their activation be modulated as required, maintaining a trade-off between performance and visibility.

2.15.3.2 The Advantage of Assertions

Assertions are considered a part of the testbench, but they are intended to follow the design. Assertions are mostly low impact and can be turned off or on as needed. It is highly recommend that assertions be turned **on** by default, at least during the early stages of integration.

Many assertions monitor initial conditions and input assumptions. Such assertions should always remain active, otherwise important integration validation will not be performed and the verification that is performed during IP-level testing cannot be relied on (as the input behavior may be different). Having an assertion *fire* (produce an error message) can save a lot of time and effort in debugging the interface.

2.15.3.3 Analog Input Sources

Some IP interfaces may end up as SoC-level interfaces, as the physical access layer to a wired or wireless medium, for example. The consideration of whether to re-use the IP-level source or create a new one at the SoC level depends on the expected functionality. As the SoC can have much more functionality, it may require a more sophisticated source.

Still, it might make sense to reuse, and possibly enhance, the IP-level signal source. This could be an easy first step for bringing up the SoC environment, to be later enhanced or augmented by other sources. Because the IP level uses constrained random control over generated signal properties, it is relatively easy to tune to the needs of the specific integration target.

2.16 UVM-MS Universal Verification Blocks

Cadence has a library of verification blocks to simplify the creation of UVM-MS test environments. The blocks carry out common tasks needed for analog verification. Using the library helps streamline the verification environment, helping you keep the architecture uniform while spending less time on low-level details. The blocks are provided with open source, such that they can be used as templates for the creation of similar verification blocks.

This section describes the verification blocks including their purpose, architecture, integration requirements, and extensions, if any.

2.16.1 Wire Verification Component

2.16.1.1 Purpose

The *wire* block contains a programmable sine wave generator and a monitor that can measure the properties of sine waves. The controlled and measured properties are: frequency, DC bias, amplitude, and phase.

Figure 2-25 Properties Controlled and Measured by the Wire UVC

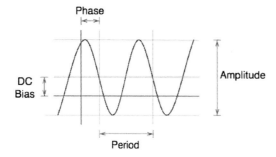

2.16.1.2 Architecture

The wire UVC implementation is split between *e* and Verilog AMS. Higher-level generation and monitoring functions are implemented in *e*, while highly active functions are implemented in Verilog AMS.

The signal generation function is driven by parameters that define the signal properties, as well as the sampling frequency. These parameters are set through a sequence item interface for a natural integration into a top-level sequence. The actual signal source is implemented in Verilog twice: once as a real-number model outputting a quantized (sampled) signal, and again as an analog block outputting a continuous wave. The selection between the two modes is done through a define macro. This selection does not affect the *e* part.

The monitor part of the block is implemented by a sampling loop implemented in Verilog as a real-number model. A parameterized sampling period is provided and is assumed to be longer than the period of the sampled signal. During that period, the sampling loop converges on the DC bias of the signal. An initial estimation of the signal frequency is derived based on zero crossings. At that time, a secondary loop is activated, performing a 20x oversampling for a duration greater than one period to accurately measure the signal properties. At the end of the measurement period, the properties are passed onto the *e* monitor for communication and coverage collection.

Signal monitoring can be triggered by an event, passed through an event port, or by a special sequence item. Monitoring is performed for a programmable period only.

Figure 2-26 Structural Diagram of the Wire UVC

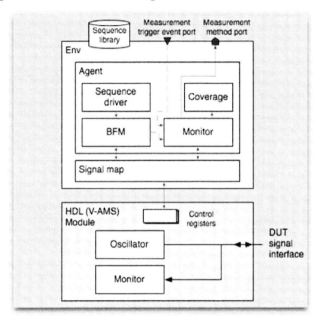

2.16.1.3 Integration

Integrating the wire UVC into a verification environment requires the following steps:

1. Instantiate the Verilog module of the wire UVC at the HDL testbench and connect the interface as appropriate. Note that the Verilog module has two interface versions: one for a single wire interface and the other for a differential pair (with a common mode tap).

2. Instantiate the *e* unit of the wire UVC at the top-level environment unit. Set the *HDL Path* of the unit to the path of the Verilog module instantiated in step 1.

3. Add constraints to configure the behavior of the generation and monitoring functions. See Example 2–35 on page 74 for more information.

4. Connect the measurement method port to the consumer of monitor results.

5. If monitoring is to be triggered by event, connect the trigger event input port. (This step is optional.)

6. Add the wire driver to the top-level sequence, so that sequence items can be activated on it.

The following three examples illustrate the integration steps.

Example 2–35   Verilog Instantiation of Wire Modules, Both Single Wire and Differential Pair Interfaces—Step 1 in the Integration Process

```
module dms_top;
  wreal osc1_p, osc1_n, cm, cm2, osc2;
  wreal aout1, aout2_p, aout2_n;
  // ---- Testbench instances --------
  dms_mswire_pair_uvc sig_gen1(.wire_c(cm),.wire_p(osc1_p), .wire_n(osc1_n));
  dms_mswire_uvc sig_gen2(.mswire(osc2));
  dms_mswire_uvc sig_mon1 (.mswire(aout1));
  dms_mswire_pair_uvc sig_mon2 (.wire_c(cm2), .wire_p(aout2_p),
.wire_n(aout2_n));
  // ---- Analog DUT -----------------
  analog_dut analog_dut(.ain1_p(osc1_p), .ain1_n(osc1_n), .ain2(osc2),
.aout1(aout1), .aout2_p(aout2_p), .aout2_n(aout2_n));
endmodule
```

Example 2–36   *e* Code Example for Instantiating, Constraining and Connecting a Wire Unit—Steps 2-5 in the Integration Process

```
unit dms_tb_env_u {
    -- Wire UVC instance
    dms_in   :dms_mswire_env is instance;
       keep dms_in.agent.active_passive == ACTIVE;
       keep dms_in.hdl_path() == "~/tb_top/sig_gen1";
       keep dms_in.clk_period == 0.50;   -- sampling clock 0.5 ns
       -- monitor portion is active as well: triggered by event
       keep dms_in.monitor_mode == EVENT_DRIVEN;
       keep dms_in.delay == 175.0;       -- delay after trigger
       keep dms_in.duration == 50;       -- clock cycles to sample

    -- Control interface instance
    vga_ctrl   :dms_reg_vga_ctrl_env is instance;
       -- changes detected in reg values trigger measurements
       -- monitor input
       keep bind(vga_ctrl.smp.change_occurred,dms_in.smp.do_sample);

    -- Connect measurement results port to checker
    checker    :dms_tb_checker_u is instance;
        keep bind(dms_in.agent.monitor.dms_mswire_transaction_complete,
              checker.dms_in_meas);
};
```

Example 2–37   *e* Code Example of Integrating a Wire Driver Into A Top-Level Sequence—Step 6 in the Integration Process

```
sequence dms_tb_sequence;      -- Define the virtual sequence

extend dms_tb_sequence {
    !dms_wire_seq    :dms_mswire_sequence; -- enable "do dms_wire_seq"
```

```
    };

    -- Add references to lower level drivers
    extend dms_tb_sequence_driver {
        dms_in    :dms_mswire_driver_u;
    };

    extend dms_tb_env_u {
        -- instantiate the virtual sequence driver
        driver :dms_tb_sequence_driver is instance;
            keep driver.dms_in == dms_in.agent.driver;
    };
```

Table 2-2 describes the wire block control parameters. These parameters should be set up as part of the instantiation, typically through constraints. While they can be changed dynamically, this usage is not recommended.

Table 2-2   Configuration Parameters for the Wire UVC

Parameter	Setting	Notes
clk_period	Constrain or assign to the sampling rate value in nSec.	Must be set. Ensure clocking is consistent for the whole testbench.
delay	A real number specifying the amount of time to wait from triggering to starting the measurement period.	Time expressed according to the time scale. Default is zero delay (measure immediately).
duration	The number of clk_period cycles to measure.	Must be set. Ensure at least two full cycles of the signal are measured.
hdl_path()	Constrain to point to the Verilog module implementing the block.	
active_passive	Constrain to either active or passive.	active by default. In active mode both signal source and monitor are active. In PASSIVE mode only the monitor is active.
monitor mode	Constrain to either event_driven or trans_driven.	event_driven by default. Determines how the monitor is triggered.
has_coverage	Constrain to either true or false.	true by default. When true, coverage is collected.

2.16.1.4 Usage

The wire block can be invoked using a transaction interface. A higher level wrapper is provided by a sequence library. Table 2-3 lists the sequence items and their fields.

Table 2-3 Sequence Items Provided for the Wire UVC

Item	Field	Value	Notes
drive_seq	ampl	Real	Peak to peak amplitude
	bias	Real	DC bias. 0 by default.
	freq	Real	Signal frequency
	phase	Real	Phase shift in degrees. 0 by default
measure_seq	delay	Real	Time delay, in units based on time scale in effect.
	duration	Integer	Number of samples to collect during measurement.

An example of driving wire UVC transactions is provided Example 2-38.

Example 2-38 *e* Code Example of A Top-Level Sequence Driving A Wire UVC Source and Monitor.

```
extend MAIN dms_tb_sequence {
    body() @driver.clock is only {
        -- setting up the signal source called osc
        do DRIVE_SEQ dms_sequence on driver.osc keeping {
            .ampl == 2.0;      -- Volts
            .bias == -5.5;     -- Volts
            .freq == 10.0e6;   -- 10 MHz
            .phase == 0.0;
        };
        -- triggering measurement on an instance called mon
        do MEASURE_SEQ dms_sequence on driver.mon keeping {
            .delay == 40.0;    -- 40 ns wait before measurement
            .duration == 200;  -- collect 200 samples of clk_period
        };
        wait [100];
        stop_run();
    };
};
```

The wire monitor reports results through a method port interface. Example 2-39 on page 77 illustrates connecting to the method port and using the measured values.

Example 2–39 *e* **Code Showing the Implementation of an Input Method Port for Reading Measured Wire Data**

```
unit dms_tb_checker_u {
    -- method port connecting to monitor
    dms_in_meas : in method_port of
        dms_mswire_measurement_done_t is instance;
    event in_meas;

    dms_in_meas(rec :dms_mswire_measurement_t) is {
        -- called each time the monitor measures a value
        out("DMS IN values:",
            "\n\tAmplitude =", rec.ampl,
            "\n\tBias      =", rec.bias,
            "\n\tPhase     =", rec.phase,
            "\n\tFrequency =", rec.freq);
        emit in_meas;
    };
};
```

The wire block collects coverage per each triggering of its monitor, if so configured. The default cover group code is provided in Example 2–40.

Example 2–40 *e* **Code of the Default Cover Group Collected By the Wire Monitor**

```
extend has_coverage dms_mswire_monitor {
    cover cov_dms_mswire_measurement is {
        item ampl : real = dms_mswire_measurement.ampl using
            ranges = {
                range([0..0.001],"mV");
                range([0.001..0.1],"10mV");
                range([0.1..0.9],"100mV");
                range([0.9..1.4],"0.9-1.4V");
                range([1.4..1.9],"1.4-1.9V");
                range([1.9..5],"1.9-5V");
                range([5..10],"5-10V");
                range([10..10e5],"10V+");
            };
        item freq : real = dms_mswire_measurement.freq using
            ranges = {
                range([0..0.9],"DC");
                range([1..10e3],"Hz");
                range([10e3..10e6],"KHz");
                range([10e6..10e7],"MHz");
                range([10e7..10e8],"10MHz");
                range([10e8..10e9],"100MHz");
                range([10e9..10e10],"GHz");
            };
    };
};
```

2.16.1.5 Extensions

The basic wire block can be extended to collect more specific coverage, adjust ranges etc. Such extensions will apply to all instances of the monitor. In order to collect coverage for specific instances, it is better to receive the measurement data through the method port provided and define specific cover groups at the receptor.

2.16.2 Simple register UVC

2.16.2.1 Purpose

The *dms_register* UVC is a simple register package specifically designed for driving the control registers of mixed-signal designs. It provides simple read/write functionality to and from a set of registers comprising an interface. An event alerts the testbench upon a value change to any of the interfaces registers.

In contrast with the full-featured *vr_ad* register package, this UVC is simple and is provided with open-source. It lacks many of the sophisticated features available in the full-featured package.

2.16.2.2 Architecture

The *dms_register* UVC features a macro generator that generates a customized UVC per *interface*. The term *interface* is used here to refer to a set of registers serving a common purpose and driven by the same block. All registers of an interface must be instantiated under the same HDL module and controlled (read and written) by the same clocking signal.

In an *e* environment, the generation and usage of specialized register UVCs is done seamlessly, in a single generate-and-run step.

Each customized UVC is structured as illustrated in Figure 2-27 on page 78.

Figure 2-27 Structure Diagram of a Generated Register UVC

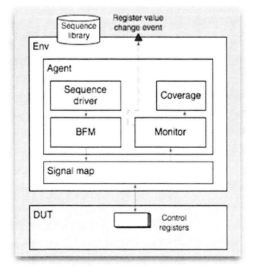

2.16.2.3 Integration

You can add a control register interface to a verification environment by following the steps below.

1. Define the interface using the *dms_reg_interface* construct. Use the same register names as used in the HDL. All registers of an interface must reside in the same HDL module (have the same path prefix).

2. Instantiate the generated interface unit within the top-level verification environment. Constrain the `hdl_path()` attribute of the unit to the module that instantiates all interface registers.

3. Add the interface driver to the top-level sequence, so that sequence items can be activated on it.

Defining register interfaces is done using the *dms_reg_interface* custom construct, which is provided by the UVC. The interface name defines the name of the generated UVC (the generated UVC environment name will be `dms_reg_intf_name_env`). The clock that is specified controls read operations, while write operations are timed by the timing of write transactions. For each register, the actual HDL register name must be provided. That name becomes the logical name of that register as well. The size, in bits and reset value, must also be specified. The bit ordering will match the HDL definition. The syntax of this construct is shown in Example 2–41.

Example 2–41 Syntax of Interface Definition Used By the dms_register UVC

```
dms_reg_interface intf_name clock intf_clock active [rise|fall] with {
  reg register_name size size_in_bits reset_val value;
  ...
};
```

Once an interface is defined, the resulting UVC environment has to be instantiated in the top-level environment. The following is an example of instantiating the specialized UVC, which is defined above.

The code listing in Example 2–42 is an example of the UVC usage.

Example 2–42 *e* Code Example Defining an Interface With Two Registers—Step 1 in the Integration Process

```
dms_reg_interface ctrl_vga clock ck_rd active rise with {
   reg ctrl_reg_a         size 4 reset_val 0b0001;
   reg ctrl_reg_b         size 8 reset_val 0b00000001;
};
```

Once an interface is defined, the resulting UVC environment has to be instantiated in the top-level environment. Example 2–43 is an example of instantiating the specialized UVC defined above.

Example 2–43 *e* **Code Example Instantiating a Control Register Interface and Connecting It—Step 2 in the Integration Process**

```
extend dms_tb_env_u {
    ctrl_vga   :dms_reg_ctrl_vga_env is instance;
        keep ctrl_vga.agent.active_passive == ACTIVE;
        keep ctrl_vga.hdl_path() == "~/tb_top/vga_d";
        -- changes detected in reg values trigger measurements monitor input
        keep bind(ctrl_vga.smp.change_occurred,dms_in.smp.do_sample);
};
```

A driver for the control interface has to be added to the top-level sequence. An example of integrating the customized UVC is shown in Example 2–44.

Example 2–44 *e* **Code Example Connecting the Control Interface Driver to the Top-Level Sequence Driver—Step 3 in the Integration Process**

```
sequence dms_tb_sequence;        -- Define the virtual sequence

extend dms_tb_sequence {
    !dms_reg_seq    :dms_reg_sequence; -- enable "do dms_reg_seq"
};
-- Add references to lower level drivers
extend dms_tb_sequence_driver {
    ctrl_vga_d   :dms_reg_driver_u;  -- a driver for each interface!
};
extend dms_tb_env_u {
    -- instantiate the virtual sequence driver
    driver :dms_tb_sequence_driver is instance;
        keep driver.ctrl_vga_d == ctrl_vga.agent.driver;
};
```

2.16.2.4 Usage

Control registers are written using a sequence-item interface. Table 2-4 summarizes the sequence items.

Table 2-4 dms_register UVC Sequence Item Parameters

Item	Field	Value	Notes
write	"name"	String	The name of the register. **Note:** string quotes are required.
	data	Integer	The size, in bits, is determined by definition
reset			

Example 2–45 on page 81 shows an example of setting up several registers of an interface by executing top-level sequence items.

Example 2–45 *e* Code Example Of Top-Level Sequence Configuring Control Registers Using Sequence Items

```
extend DEF_SETUP dms_tb_sequence {
    control_val_1  :uint (bits:8);
       keep soft control_val_1 == 0xF0;
    control_val_2  :uint (bits:3);
       keep soft control_val_1 == 0xF0;
    body() @driver.clock is only {
       message(HIGH,"Control register set up\n",
               "\tcontrol_reg_1 = ", hex(control_val_1),"\n",
               "\tcontrol_reg_2 = ", hex(control_val_2),"\n",
              );
       do WRITE dms_ctrl_seq on driver.ctrl_vga_d keeping {
          .reg_name == "ctrl_reg_1";
          .data == control_val_1;
       };
       do WRITE dms_ctrl_seq on driver.ctrl_vga_d keeping {
          .reg_name == "ctrl_reg_2";
          .data == control_val_2;
       };
    };
};
```

Control registers can be read by using the method API in Example 2–46, sent to a customized dms_register environment. This method of *back-door* reading is useful when the testbench needs to access the register value without interfering with execution.

Example 2–46 *e* Code Back-Door API For Accessing Register Values

```
dms_reg_env.get_val(reg_name :string):int (bits:*)
```

Back-door access to register values is sometime used for collecting coverage. An example of this is shown in Example 2–47.

Example 2–47 *e* Code Example Using Back-Door Access Method To Register Values

```
unit ctrl_monitor_u {
    ctrl_vga     :dms_reg_ctrl_vga_env; -- reference to register package
    regs_changed :in event_port is instance;
    event config_change is @regs_changed$;
    cover config_change is {
       -- both items read through back door API
       item ctrl_reg_1: int (bits:8) = ctrl_vga.get_val("ctrl_reg_1")
          using illegal = (ctrl_reg_1 == 0xFF);
       item ctrl_reg_2: int (bits:3) = ctrl_vga.get_val("ctrl_reg_2");
    };
};
extend dms_tb_env_u {
    -- instantiate the monitor
```

```
        ctrl_mon : ctrl_monitor_u is instance;
        keep ctrl_mon.ctrl_vga == ctrl_vga;
        -- connect the register change event to the monitor
        keep bind(ctrl_vga.smp.change_occurred,ctrl_mon.regs_changed);
    };
```

Note The dms_register UVC collects coverage on each register by default. Coverage collection can be disabled by constraining `has_coverage == false` for the instantiated interface.

2.16.3 Analog to Digital Converter (ADC) UVC

The *dms_adc* UVC is a block that performs a conversion from a real value to a digital value. The number of bits of the digital output is parameterized. Coverage is collected, but no generation or checking functions are provided.

2.16.3.1 Architecture

The ADC block contains a monitor and a signal map enclosed in an agent unit. The monitor is driven by an input sample, which results in a converted value at the output. Input is provided by the connection of simple ports to either HDL signals or other *e* units.

The signal map needs to connect to the digital input. It also needs to provide a trigger event and a reference voltage (the voltage that would yield the highest reading at the output).

Each time the ADC is invoked, a zero-time conversion is performed. The results are provided through a method port. The called method is expected to be a monitor or scoreboard function, but can easily be made to drive an HDL signal with the result, if desired.

2.16.3.2 Integration

Integrating the ADC block can be done in two ways: a simple port connection or a method port connection.

For a **simple port input** connection do the following:

1. Instantiate the ADC block within the environment. Constrain the output width in bits by setting the parameter *adc_bits*.

2. Connect the following ports in the ADC signal map:

    - *port_ana_in*—Connect to a real value to be converted.
    - *port_v_ref*—Connect to a constant real value representing the maximum converted value.
    - *port_do_sample*—A binary value that triggers the measurement on a *rise* transition.

3. Connect the output method port *dms_adc_sample* to a compatible method for handling the output.

Analog to Digital Converter (ADC) UVC

For a **method port input** connection do the following:

1. Instantiate the ADC block within the environment. Constrain the output width in bits by setting the parameter *adc_bits*.

2. Connect the following ports in the ADC signal map:

   - *dms_adc_input*—Connect to a compatible method port of type *dms_adc_input_t*.
   - *port_v_ref*—Connect to a constant real value representing the maximum converted value

3. Connect the output method port *dms_adc_sample* to a compatible method for handling the output.

An example instantiation of the ADC block is listed in Example 2–48.

Example 2–48   Instantiating the ADC Block Using Simple Port Connections

The ADC Output Size Is Set To 12 in this example.

```
unit dms_tb_env_u like any_unit {
    -- instantiate the ADC UVC
    adc_1 :dms_adc_env is instance;
        keep adc_1.smp.port_do_sample.hdl_path() == "~/dms_top/do_sample";
        keep adc_1.smp.port_ana_in.hdl_path()== "~/dms_top/adc_out";
        keep adc_1.smp.port_v_ref.hdl_path()== "~/dms_top/v_ref";
        keep adc_1.adc_bits == 12;

    event clk is change('~/dms_top/sig_clock')@sim;
};
```

The listing in Example 2–49 is an example of connecting the ADC output port.

Example 2–49   *e* Code Example Using ADC Block Output In A Monitor

```
unit dms_tb_monitor {
    -- incoming samples from TB ADC
    adc_sample : in method_port of dms_adc_sample_t is instance;
    adc_sample(val :uint) is {
        -- this method is called each time the ADC returns a sample
        out("Monitor ADC returned: out=", val);
    };
};

extend dms_tb_env_u {
    adc_mon   : dms_tb_monitor is instance;
        -- port binding
        keep bind(adc_mon.adc_sample, adc_1.smp.dms_adc_sample);
};
```

2.16.3.3 Usage

Note that the ADC block collects coverage on each sample by default. Coverage collection can be disabled by constraining `has_coverage == false` for the instantiated interface. The cover group ranges can be adjusted by extending the cover group called `cov_dms_adc_measurement`.

2.16.4 Digital to Analog Converter (DAC) UVC

2.16.4.1 Purpose

The *dms_dac* UVC is a block that performs conversion from a digital vector to a real value. The number of bits of the digital input is parameterized. Coverage is collected, but no generation or checking functions are provided.

2.16.4.2 Architecture

The DAC block contains a monitor and a signal map enclosed in an agent unit. The monitor is driven by an input sample, which results in a converted value at the output. Input can be provided by connecting to HDL signals, as well as through an *e* method port.

When connected directly to HDL signals, the signal map needs to connect to the analog input. It also needs to provide a trigger event and a reference voltage (the voltage that would yield the highest reading at the output).

When connecting through the *e* method port, no trigger event is needed, but the reference voltage must still be provided.

Each time the ADC is invoked, either by the trigger event or the method port, a zero-time conversion is performed. The results are provided through a method port. The called method is expected to be a monitor or scoreboard function, but can easily be made to drive an HDL signal with the result, if desired.

2.16.4.3 Integration

To integrate the DAC block, perform the following steps:

1. Instantiate the DAC block within the environment. Constrain the output width in bits by setting the parameter *dac_bit*s.

2. Connect the following ports in the DAC signal map:

    - *port_dig_in*—Connected to a vector of bits to be converted.
    - *port_v_ref*—Connected to a constant real value representing the maximum converted value.
    - *port_do_sample*—A binary value that triggers the conversion on a *rise* transition.

3. Connect the output method port *dms_dac_sample* to a compatible method for handling the output.

The listing in Example 2–50 shows the instantiation of the DAC block.

Example 2–50 Instantiating the DAC Block, Using Simple Port Connections

The DAC Input Size Is Set To 8 in this example.

```
unit dms_tb_env_u like any_unit {
    -- instantiate the UVCs
    dac_2 :dms_dac_env is instance;
        keep dac_2.smp.port_sample.hdl_path()== "~/dms_top/do_sample";
        keep dac_2.smp.port_dig_in.hdl_path()== "~/dms_top/dac_out";
        keep dac_2.smp.port_v_ref.hdl_path()== "~/dms_top/v_ref";
        keep dac_2.dac_bits == 8;
    event clk is change('~/dms_top/sig_clock')@sim;
};
```

Example 2–51 shows an example of connecting the DAC output port.

Example 2–51 *e* Code Example Using DAC Block Output In A Monitor

```
unit dms_tb_monitor {
    -- incoming samples from TB ADC
    dac_sample : in method_port of dms_dac_sample_t is instance;
    dac_sample(val :real) is {
        -- this method is called each time the DAC returns a sample
        out("Monitor DAC returned: out=", val);
    };
};

extend dms_tb_env_u {
    dac_mon   : dms_tb_monitor is instance;
        keep dac_mon.v_ref == v_ref;
        -- port binding
        keep bind(dac_mon.dac_sample, dac_2.smp.dms_dac_sample);
};
```

2.16.4.4 Usage

Be aware that the DAC block generates positive values only, the input is assumed to be an unsigned vector.

Also note that by default, the DAC block collects coverage on each output. Coverage collection can be disabled by constraining `has_coverage == false` for the instantiated interface. The cover group ranges can be adjusted by extending the cover group called `cov_dms_dac_measurement`.

2.16.5 Level Crossing Monitor

2.16.5.1 Purpose

The *dms_threshold* UVC is a block that monitors a real value level with respect to a reference value. It triggers when a threshold crossing is detected. Both *time* and *value* tolerances can be specified to make the monitor ignore small signal changes or rapid spikes.

2.16.5.2 Architecture

The threshold block is implemented partly in *e* and partly in Verilog AMS. This is done for performance reasons, because the continuous monitoring of a value is compute intensive. The *e* portion contains a monitor and a signal map enclosed in an agent unit. Constraints applied at the UVC top-level are passed on to the Verilog-AMS part upon initialization.

The low-level Verilog-AMS block is connected to the monitored signal and a reference value, which is not assumed to be constant.

Each time the monitor detects a threshold crossing that is longer than the minimal time window, and greater than the minimal value change, the crossing is reported. The following mechanisms are used for reporting:

- Three event ports present on the signal map interface may be triggered:
  - A low-to-high transition event.
  - A high-to-low transition event.
  - A generic transition event.
- A method port present on the signal map is called. The port passes a parameter indicating the kind of detected crossing.

The monitor collects coverage, if configured to do so. Coverage includes the occurrence of crossings and the direction.

2.16.5.3 Integration

Integrating the threshold detector block requires that you perform the following steps:

1. Instantiate the Verilog module of the threshold UVC in the HDL testbench and connect the interface as appropriate.

2. Instantiate the *e* unit of the threshold UVC in the top-level environment unit. Set the HDL path of the unit to the path of the Verilog module instantiated in step 1.

3. Add constraints to configure the behavior of the threshold monitor. See Example 2–52 for more information.

4. Connect the measurement method port to the consumer of the monitor results or connect the event ports that are reporting crossings, as desired.

Level Crossing Monitor

The following three examples illustrate the integration steps.

Example 2-52 Verilog Instantiation Of the Threshold Monitor—Step 1 in the Integration Process

```
module dms_top;
   wreal sig1;
   wreal v_ref
   // ---- Testbench instances --------
   dms_threshold_uvc  mon(.signal(sig1), .v_ref(v_ref));

   // ---- Analog DUT -----------------
   analog_dut analog_dut( .aout(sig1));

   initial
      v_ref = 0.9; // setting a reference value
endmodule
```

Example 2-53 *e* Code Example For Instantiating and Constraining a Threshold Unit—Steps 2 and 3 in the Integration Process

```
unit dms_tb_env_u like any_unit {
    resolution :real; -- sampling clock resolution, centrally controlled
    keep resolution == 2.0;
    -- instantiate the UVCs
    mon_1 :dms_threshold_env is instance;
       keep mon_1.smp.hdl_path()== "~/dms_top/mon";
       keep mon_1.clk_period == resolution;
       keep mon_1.monitor_min_time == 50.0;   -- ignore spikes < 50 ns
       keep mon_1.monitor_min_voltage == 0.2; -- voltage margin is +/- 0.2 V

    event clk is change('~/dms_top/sig_clock')@sim;
};
```

Example 2-54 *e* Code Hooking Up the Monitor Event Port Output—Step 4 in the Integration Process

```
unit dms_tb_mon {
    rise_crossing :in event_port is instance;
    event cross_detected is @rise_crossing$;
    on cross_detected {
       out("Crossing from low to high detected");
    };
};

extend dms_tb_env_u {
    cross_reporter :dms_tb_mon is instance;
       keep bind(mon_1.smp.port_l_h_cross, cross_reporter.rise_crossing);
};
```

Advanced Verification Topics

Table 2-5 specifies the threshold UVC control parameters. These parameters should be set up as part of the instantiation, typically through constraints. Changing these parameters during runtime has no effect and is therefore discouraged.

Table 2-5 Configuration Parameters for the Threshold Monitor UVC

Parameter	Setting	Notes
clk_period	A value in ns. Constrain or assign this to the sampling rate.	Must be set. Ensure clocking is consistent for the entire testbench.
monitor_min_time	A real number measured in ns. The monitor ignores changes shorter than this value.	Must be set.
monitor_min_voltage	A real number. Voltage variations smaller than this value are ignored by the monitor.	Must be set.
hdl_path()	Constrain to point to the Verilog module implementing the block.	
has_coverage	Constrain to true or false.	true by default. When true, coverage is collected.

Table 2-6 summarizes the port connectivity available at the interface of the threshold UVC signal map.

Table 2-6 Available Port Connections for Integrating the Threshold Monitor

Port Name	Kind	Function
dms_threshold_measurement	Output method port of type dms_measurement_threshold_done_t, passing a single parameter of type dms_threshold_stat_t	Passes one of the following values: FALLING, RISING, or UNCERTAIN
port_h_l_cross	Output event port.	Triggers when high-to-low crossing is detected (Falling)
port_l_h_cross	Output event port.	Triggers when low to high crossing is detected (Rising)
port_cross	Output event port.	Triggers each time a crossing is detected

2.16.6 Ramp Generator and Monitor

2.16.6.1 Purpose

The *dms_ramp* UVC generates and monitors arbitrary piece-wise wave shapes. It can be used to create particular a-periodic wave forms at a relatively low frequency, using sequence items. A typical use would be emulating a noisy power supply, for example.

Figure 2-28 Voltage and Time Points Controlled and Measured by the Ramp UVC

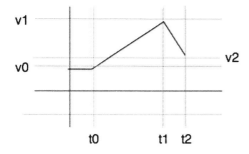

2.16.6.2 Architecture

The *ramp* UVC implementation is split between *e* and Verilog AMS. While generation functions are implemented in *e*, some highly active monitoring functions are implemented in Verilog AMS.

The generation function is sequence driven, where each transaction sets the next time-voltage goal.

The monitor part of the block is implemented by a sampling loop written in Verilog as a real-number model. A parameterized sampling period is provided, and is assumed to be much faster than the fastest transient. When the monitor detects a change in the time derivative (an elbow, or a change in direction), the current linear segment is reported and the tracking of the new linear segment is started from the last elbow.

Figure 2-29 Structural Diagram of the Ramp UVC

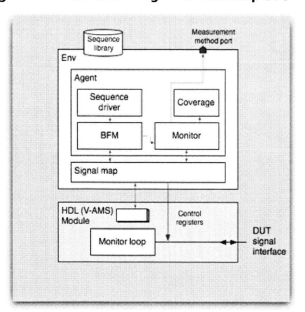

2.16.6.3 Integration

Integrating the ramp UVC into a verification environment requires the following steps:

1. Instantiate the Verilog module of the ramp UVC at the HDL testbench and connect the interface as appropriate.

2. Instantiate the *e* unit of the ramp UVC at the top-level environment unit. Set the HDL Path of the unit to the path of the Verilog module instantiated in step 1.

3. Add constraints to configure the behavior of the generation and monitoring functions. See Example 2–55 for more information.

4. Connect the measurement method port to consumer of monitor results

5. Add the ramp driver to the top-level sequence, so that sequence items can be activated on it.

The following four code examples illustrate the integration steps.

Example 2-55 Verilog Instantiation Of Ramp Modules, One For Driving and One For Monitoring the Output—Step 1 in the Integration Process

```
module dms_top;
  wreal sig1;
  wreal sig2
  // ---- Testbench instances --------
  dms_ramp_uvc  gen(.mswire(sig1));
```

```
  dms_ramp_uvc  mon(.mswire(sig2));

  // ---- Analog DUT -----------------
  analog_dut analog_dut(.ain(sig1), aout(sig2));
endmodule
```

Example 2–56 *e* **Code Example for Instantiating, Constraining and Connecting the Ramp UVCs—Steps 2 and 3 in the Integration Process**

```
unit dms_tb_env_u like any_unit {
    resolution :real; -- sampling clock resolution, centrally controlled
    keep resolution == 2.0;
    -- instantiate the UVCs
    osc_1 :dms_ramp_env is instance;
        keep osc_1.agent.active_passive == ACTIVE;
        keep osc_1.smp.port_sig.hdl_path() == "~/dms_top/sig1";
        keep osc_1.smp.hdl_path()== "~/dms_top/gen";
        keep osc_1.clk_period == resolution;
        keep osc_1.monitor_noise_threshold == 0.1; -- ignore 0.1 V noise
        keep osc_1.monitor_min_ramp_time == 50.0;
-- ignore ramps shorther than 50 ns
        keep osc_1.monitor_min_ramp_voltage == 0.3;
-- ignore ramps where dV < 0.3V
    mon_1 :dms_ramp_env is instance;
        keep mon_1.agent.active_passive == PASSIVE;
        keep osc_1.smp.hdl_path()== "~/dms_top/mon";
        keep osc_1.clk_period == resolution;
        keep osc_1.monitor_noise_threshold == 0.1; -- ignore 0.1 V noise
        keep osc_1.monitor_min_ramp_time == 50.0;
-- ignore ramps shorther than 50 ns
        keep osc_1.monitor_min_ramp_voltage == 0.3;
-- ignore ramps where dV < 0.3V

    event clk is change('~/dms_top/sig_clock')@sim;
};
```

Example 2–57 *e* **Code Hooking Up the Monitor Output—Step 4 in the Integration Process**

```
unit dms_tb_mon {
    meas_port :in method_port of dms_ramp_measurement_done_t is instance;
    meas_port(val :dms_ramp_measurement_t) is {
        -- Reporting code, scoreboard etc.goes here
        out("Ramp started at ", val.t_start, " ended at ", val.t_end,
            " slope = ", (val.v_end-val.v_start)/(val.t_end - val.t_start));
    };
};

extend dms_tb_env_u {
    reporter : dms_tb_mon is instance;
        keep bind(reporter.meas_port, osc_1.smp.dms_ramp_measurement);
};
```

Example 2–58 *e* **Code Example Of Integrating a Ramp Driver Into a Top-Level Sequence—Step 5 in the Integration Process**

```
sequence dms_tb_sequence;    -- The main virtual sequence
extend dms_tb_sequence {
    !ramp_sequence      :dms_ramp_sequence;    -- add the ramp kind
};
-- Reference the instantiated drivers
extend dms_tb_sequence_driver {
    osc_1 :dms_ramp_driver_u;

extend dms_tb_env_u {
    -- instantiate the virtual sequence driver
    driver :dms_tb_sequence_driver is instance;
        keep osc_1.agent is a ACTIVE dms_ramp_agent (a_agent) =>
            driver.osc_1 == a_agent.driver;
};
```

Adding the *ramp* sequence driver to the virtual sequence allows direct control of the generated output from the main sequence.

Table 2-7 specifies the *ramp* UVC control parameters. These parameters should be set up as part of the instantiation, typically through constraints. While they can be changed dynamically, such usage is not recommended.

Table 2-7   Configuration Parameters for the Ramp UVC

Parameter	Setting	Notes
`clk_period`	A value in ns. Constrain or assign this to the sampling rate.	Must be set. Ensure clocking is consistent for the entire testbench.
`monitor_noise_threshold`	A real number. Voltage changes smaller than this parameter are ignored (considered noise).	Default is 0. This value determines how fine the piece-wise linear approximation is.
`min_ramp_time`	A real number, measured in ns. The monitor ignores ramps shorter than this value.	Must be set.
`min_ramp_voltage`	A real number. Voltage variations smaller than this value are ignored by monitor.	Must be set.
`hdl_path()`	Constrain to point to the Verilog module implementing the block.	

Table 2-7 Configuration Parameters for the Ramp UVC (continued)

`active_passive`	Constrain to either `active` or `passive`.	`active` by default. In `active` mode both signal source and monitor are active. In `PASSIVE` mode only the monitor is active.
`has_coverage`	Constrain to either `true` or `false`.	`true` by default. When `true`, coverage is collected.

2.16.6.4 Usage

The ramp can be invoked using a transaction interface. A higher level wrapper is provided by a sequence library. Table 2-8 lists the sequence item and its fields.

Table 2-8 Sequence Items Provided for Ramp UVC

Item	Field	Value	Notes
`TRANS_RAMP_SEQ`	voltage	Real	The voltage end point for the next segment.
	duration	Real	The time duration of the next segment.

Be aware that the sequence uses the end point of the previous sequence as its starting point.

Example 2–59 shows an example of driving ramp UVC transactions.

Example 2–59 *e* Code Example of a Top-Level Sequence Driving a Wire UVC Source and Monitor

```
extend MAIN dms_tb_sequence {
    body() @driver.clock is only {
        wait[10]; -- clock cycles
        out("Osc1 ramp up");
        do TRANS_RAMP_SEQ dms_sequence on driver.osc_1 keeping {
            .voltage == 1.8;    -- Target voltage
            .duration == 300.0; -- ns
        };
        wait [4];
        out("Osc1 ramp down");
        do TRANS_RAMP_SEQ dms_sequence on driver.osc_1 keeping {
            .voltage == 0.1;
            .duration == 300.0; -- ns
        };
        wait [10];
        out("Osc1 fast ramp up - should be filtered");
```

```
            do TRANS_RAMP_SEQ dms_sequence on driver.osc_1 keeping {
                .voltage == 1.8;
                .duration == 35.0; -- ns
            };
            wait [3];
            out("Osc1 ramp down to 0.9 V");
            do TRANS_RAMP_SEQ dms_sequence on driver.osc_1 keeping {
                .voltage == 0.9;
                .duration == 200.0; -- ns
            };
            wait [3];
            out("Osc1 limited ramp down (0.7V - should be filtered");
            do TRANS_RAMP_SEQ dms_sequence on driver.osc_1 keeping {
                .voltage == 0.7;
                .duration == 200.0; -- ns
            };
            wait [100];
            out("Done");
            stop_run();
        };
    };
```

Executing the sequence in Example 2-59 on page 93 results in the waveform shown in Figure 2-30.

Figure 2-30 Ramp Waveform Resulting Generated by Above Sequence

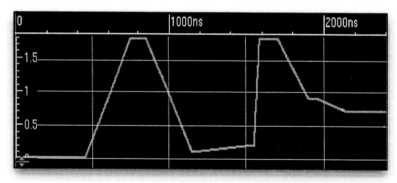

The ramp monitor provides coverage by default. Coverage is sampled on each linear segment, covering the voltage range and slew range.

2.17 Summary

Convergence, miniaturization, and mobility are driving more analog content into our digital world. An obvious measure is the array of more sophisticated analog circuits supporting high-speed communications, touch screens, high-fidelity audio, and other features. These are designs that are both high in performance and have complex signal paths through both analog and digital components. This era of process technology has also allowed analog and mixed-signal designers to begin to integrate significant amounts of the functionality of the entire systems onto a single chip. The more subtle aspect is the mutual dependency of analog and digital circuits now needed to deliver these product features. If analog development could continue on a separate, parallel path with digital, then projects could continue to succeed simply by applying additional analog engineering resources coupled with incremental changes in methodology. You are reading this because you know that's not the case.

Most integrated circuits developed today are mixed-signal; an analog chip with some digital content, or a digital chip with some analog content. However, analog mixed-signal must evolve to be digital-mixed signal in our digital world. The first step is to expand the digital concept of a verification plan to analog. The verification plan defines the functionality to be verified, and with product features that have digital functionality that depends on analog circuits, the current notion of planning must expand. The next logical step is to raise analog simulation performance through model abstraction. While behavioral standards like Verilog-A, Verilog-AMS, and VHDL-A do exist, the required performance can only be achieved by modeling these analog blocks using real-number models (in a digital context) along with real-number randomization, assertions, and functional coverage. This would provide significant performance improvements because the entire mixed-signal chip verification can be performed at digital speeds. The IEEE 1647 *e* language already has this processing, and a similar set of capabilities has been proposed for IEEE 1800 SystemVerilog at the time this chapter was written. These two steps allow UVM and metric-driven verification to now extend to cover mixed-signal verification.

While this change may appear evolutionary to digital verification engineers, it is revolutionary to most analog engineers. As such, project managers should begin planning for these methodology changes through pilot projects and engineering training. It may be a digital device, but it is truly a mixed-signal world.

3   Low-Power Verification with the UVM

This chapter provides you with an introduction to low-power verification and references the UVM for the verification environment for low-power (LP) designs. It contains the following topics:

- Low-power verification methodology overview
- Understanding low-power design and verification challenges
- Low-power discovery and verification planning
- Creating a power-aware uvm environment
    - Understanding the low power universal verification component (lp-uvc)
- Running low-power verification
- Common low-power issues and how to avoid them

3.1   Introduction

Low-power decisions have become a driver of design architecture and product decisions. Mobile devices are constantly trading-off the desire for new features against the requirements for battery life. Large servers and processors are facing requirements to conserve power in order to comply with strict energy standards and to improve reliability and design costs. To meet these requirements, designers are increasingly implementing advanced power-savings techniques in the hardware. The verification environment needs to adapt to include these low-power features in the verification process and to provide a comprehensive solution for the verification of these designs.

Many of today's low-power designs include techniques previously reserved for advanced users. These techniques, like power shutoff, dynamic voltage scaling, and bias modes are becoming increasingly commonplace. The expansion in the use of these advanced low-power techniques has been fuelled by the advent of power-intent languages such as the *common power format* (CPF) and the *unified power format* (UPF). These languages enable the automation of low-power design and verification by filling the abstraction gaps in the RTL design languages needed to implement the power intent. The languages require the underlying simulator to provide a silicon-accurate simulation of this low-power intent. These power-intent languages provide a consistent view of the low-power architecture throughout the verification and implementation flow.

The last point cannot be stressed enough; low power spans through multiple disciplines and goes beyond what is traditionally thought of as functional verification. The low-power verification methodology includes everything from verifying electrical correctness, to the more traditional functional verification of the design running in each power mode, all the way to verifying that the hardware and software control is correctly exercising all the modes and obtaining the expected low-power performance.

This chapter provides an overview of the low-power design and verification methodology. The focus is on the functional verification and the use of the UVM and planning-driven verification to address the challenges of low-power design.

3.1.1   What is Unique about Low-Power Verification

Low-power design provides a number of unique challenges for the verification team. Even understanding what needs to be verified can be a hurdle for verification teams new to low-power design. Traditionally, functional verification teams never had to concern themselves with electrical shorts or excessive noise and unpredictability due to power switching. But with low-power design, these and other issues become items to verify and can have assertions and checks defined to detect issues at the simulation level. Checking for proper power modes and control-signal sequences can avoid these very real issues appearing in the final chip. Fortunately, a few high-level concepts help tremendously with understanding and developing low-power verification environments.

The first concept is that RTL is no longer enough to describe the design, which is now a combination of the RTL, the power intent, and the software running on the system. The power intent contains the details of the low-power architecture that will be inferred by the implementation tools. The verification environment needs to understand and properly model these details or it is impossible to completely verify the low-power architecture. The accuracy of the model is directly related to the accuracy of the verification flow. Power intent languages like CPF and UPF are tremendous productivity enhancers, allowing the designer to optimize the power intent very quickly. The verification methodology needs to support this rapidly changing design environment.

The second principle is that low-power designs greatly increase the complexity and verification space of a design. For example, a hierarchical device with 8 IP blocks that can be powered off introduces 256 modes of operation and over 65,000 possible power transitions. A desire to have flexible control of power from software has seen some designs explode to 70,000 modes. The verification team needs to determine how to address this complexity without incurring such a drastic increase in verification cost. The full scope involves determining how to enter and exit each of these modes, knowing full well that the control of the power state can be a complex interaction between hardware and software. Furthermore, the team needs to identify exactly what needs to be tested for each of those modes of operation to prove that each operates correctly. Clearly, a systematic methodology needs to be applied to low-power verification. The UVM and metric-driven verification together provide an excellent framework for addressing this challenging and ever-changing low-power environment.

Finally, low-power verification methodology has a much broader scope than is traditionally seen. The methodology must go beyond the RTL verification and can include emulation, HW/SW co-simulation, assertion-based verification, formal analysis, planning, coverage, logical equivalency checking, and low-power design checks. The verification engineers not only have to be familiar with the low-power architectures, but also with what tools are available and best suited for a particular verification task.

3.1.2 Understanding the Scope of Low-Power Verification

Low power affects all levels of the hardware and software design for a system. The verification tasks can be separated into several high-level categories:

- **System-Level Power Architecture and Control**

  A complex combination of hardware and software is often required to fully use the power savings enabled by a large array of power modes. The low-power verification effort must ensure that the system-level architecture is utilizing the power modes efficiently and correctly. In some cases, incorrect power-mode transitions can cause a device to fail and, in extreme cases, get hot enough to melt components. Most cases are not that severe, but often lead to very poor performance and missed low-power opportunities.

- **System-Level Power Behavior**

  The interaction between different power domains in each mode needs to be verified. These interfaces are the most common source of verification problems. An incorrect interface can cause system buses to lock up, preventing domains from powering up, or can present or propagate incorrect data through the system. In particular, problems tend to occur in power-shutoff designs and between domains at different voltage levels. As the number of modes and domains increases, the complexity of this logic also increases.

- **Modeling Low-Power Structures**

  The low-power structures that are part of the power architecture specification do not exist in the RTL design. The verification tools need to correctly model these structures and provide a power-aware simulation at the RTL level and higher. They also need to ensure that what is modeled by the simulator matches what is implemented by synthesis and place and route. This is absolutely critical for power-shutoff designs, but it is also a factor in multiple-voltage and dynamic-voltage designs as well. This modeling can be automated by both static and dynamic verification tools, but the results need to contribute within the bounds the overall low-power methodology.

- **Verification of Power Control Structures**

  There is a layer of logic between the system-level power control and the modeling of the low-level power structures that initiates power-mode transitions. This logic controls underlying hardware that cycles through a sequence of events that drives the low-level structures, such as isolation, state retention and the power shutoff. Low-power verification must ensure that this sequence happens properly, that the system works properly with this domain shutoff and that its outputs are isolated, and that the device can power on properly. The methodology includes the use of automatically generated assertions to verify these control structures.

- **Implementation**

  Low-power verification needs to ensure that what was implemented matches what was verified and correctly implements the low-power constructs, and that the implementation meets all power related design guidelines. The CPF flow helps ensure this using a *correct-by-construction* approach using the same low-power specification for both verification and implementation. Power-aware formal verification tools provide a formal method for verifying these items. They can provide a rich set of design checks to ensure that the implemented design correctly implements the power intent. The advanced low-power design techniques greatly increase the complexity of this verification.

3.1.3 Low-Power Verification Methodology

A complete, low-power verification solution requires verification that goes beyond what is typically seen in the UVM. The low-power flow needs to be a holistic flow that includes low-power design checks, logical equivalency checking, and physical verification of low-power hardware. The flow diagram below shows a simplified representation of the overall verification flow for low power.

Figure 3-1 Low-Power Verification Flow Diagram

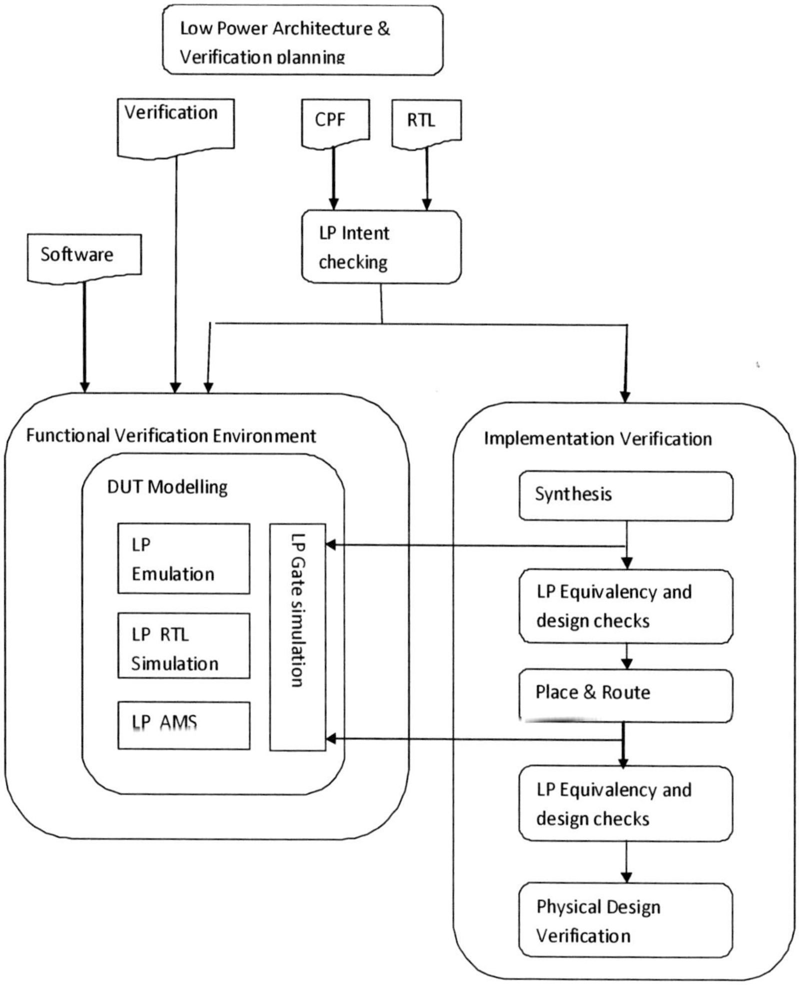

The low-power verification flow naturally starts with the definition of the low-power architecture. This definition may take advantage of electronic system level (ESL) models, SystemC synthesis, and chip-planning solutions to define and optimize the architecture. This architecture drives the power intent, the RTL, and the verification plan. The power intent and RTL are then used throughout the implementation and verification flow. ***This is key to the low-power methodology. The RTL and CPF become the new design specification.*** Ensuring that all the tools in the flow use this same power intent provides what is referred to as a *closed-loop verification flow*. It ensures that what was simulated and emulated match what is implemented. This is the cornerstone of the low-power verification flow.

Verification planning occurs in parallel with the architectural definition and plays a critical role in most low-power designs. As discussed earlier, low power can lead to a large increase in the complexity of the design. Planning is the only effective method to address that complexity and drive to verification closure.

The first step in the verification flow is to validate that the power intent is complete and correct. These checks are highly automated and use formal engines to verify a number of low-power design properties. Identifying issues before costly simulation and synthesis runs is a key productivity feature of the flow. The design checks identify missing isolation and level shifting between domains, ensure that power control itself is properly powered, and identify a host of other functional and electrical issues. The formal techniques are not only more exhaustive than simulation, they can also significantly reduce the dynamic simulation requirements. In addition, these checks are actually done on the power intent in conjunction with RTL, very early in the design cycle as opposed to traditional methods.

The overall verification environment is designed around the UVM and provides a complete metric-driven verification environment. The UVM provides the methodology used to validate the system-level power architecture and control and ensures that the system operates correctly under each of its modes of operation. The UVM is used to drive the various low-power scenarios and provide the coverage and checking to measure the results of those scenarios. In short, the UVM provides the core verification environment for the functional aspects of the low-power verification methodology.

Central to this verification environment are core engines that understand the low-power intent. Accurately modeling the low-power hardware is critical to successful verification. The modeling goes beyond simple tracking of the voltage levels in the design. It must include detailed modeling of power-shutoff corruption in complex hierarchical power architectures and mimic the implementation tools in insertion of LP cells. Nearly all low-power designs also include embedded software that provides much of the power management for the system. Power-aware emulation is often required to cover a large enough time window to verify this software. Finally, analog mixed-signal simulations are also required in LP most designs.

On the right-hand side of Figure 3-1 on page 100, the implementation flow and its verification are also critical to low-power design. As the RTL and CPF move through the implementation flow, there is a need to verify that each design transformation is complete and correct. The flow uses low-power-aware logic equivalency checking coupled with static design checks. This type of verification typically falls outside the scope of the verification team, but the lines between groups are becoming blurred. At a minimum, the RTL-level design checks should be considered as part of the front-end verification team's tasks.

The LP solution is designed to address the full breath of low-power design. The strength of the solution is the use of the same power intent throughout the flow and the use of right technology for the right task. Design checks and equivalency checking are used to provide exhaustive verification wherever possible. Power-aware simulation and emulation provide the core engines, enabling the full metric-driven verification flow required for verification closure.

3.1.4 Understanding Low-Power Verification

The goals of the UVM LP solution are simple: Provide methodology and automation to prevent bug escapes in low-power design.

What types of bugs need to be addressed? Some of them include:

- IP block powers down and then system hangs
- IP block powers off but cannot be powered back up again
- System power is higher than expected
- Electrical issues when power modes transition
- Functional errors in a block when adjacent blocks power off
- Lost packets or data
- Inconsistent results in real hardware

This is just a handful of issues that can occur when the low-power architecture is not properly understood and verified. This UVM LP methodology provides an introduction to those low-power techniques, provides guidelines for how to plan and define a reusable verification environment that is complete and can detect these issues.

3.2 Understanding Low-Power Design and Verification Challenges

Before diving into the low-power verification methodology, it is important to understand the basics of the low-power architectures and the goals for those architectures.

3.2.1 How Low-Power Implementations Are Designed Today

The first thing to understand about modern low-power verification is that the RTL alone is no longer sufficient to describe the function of the design—it also requires power intent. To perform low-power design and verification the combination of RTL and power intent is required. This power intent is provided by side files such as CPF or UPF.

The first question asked by many designers is "Why invent a new language to capture the power intent?" The answer is that the following are needed:

- Language to cover the full flow:
  - The power intent is used for everything from RTL simulation and synthesis to place and route and physical verification.
  - While RTL could be extended for the front-end tasks, it is not enough to drive the implementation flow. Details about the technology used and specifics about voltage levels are simply not appropriate for the RTL.

- Closed-loop flow
  - Having a formal language that covers the whole flow is a critical step in ensuring a consistent view of the low-power intent across the front-end and back-end design tasks.
  - It allows verification tools to validate that what was simulated is the same as what was implemented.
- Reuse scheme where the same RTL can be used for multiple power architectures
  - The side file approach enables the same RTL to be instantiated multiple times, with each instance having different power intent.
  - Capturing the low-power architecture in the RTL would hard code the power architecture into the source.

The low-power formats enabled the rapid proliferation of the more advanced low-power architectures by providing a methodology to ensure the functional correctness and the necessary design automation. On the verification side, it is critical to the closed-loop flow—critical to ensuring what was implemented by synthesis and place and route matches what was verified.

3.2.2 Challenges for Low-Power Verification

There are two major challenges for low-power verification. The first is modeling the low-power features based on the low-power intent (CPF/UPF). The second challenge is handling the increased complexity caused by a large number of power modes being added to a design. As previously mentioned, it is common to have designs with 10 to 20 unique power modes. Without a planning-driven approach to verification, this could translate into an orders-of-magnitude increase in verification effort.

Figure 3-2 Chart of Low-Power Complexity Due to Power Modes

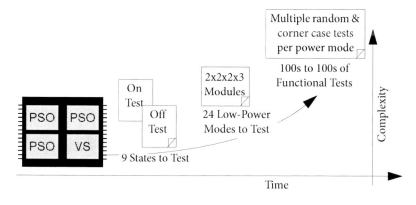

The UVM LP methodology starts with a planning-driven approach. It allows the design team to agree on the priorities and to organize the verification effort to provide maximum coverage for the design.

3.2.3 Low-Power Optimization

The goal of all these low-power techniques is to reduce either the dynamic (switching) power and/or the static (leakage) power of the design. A simple equation of the power consumption is provided for each of these in the following table.

$$P_{switching} = TR \cdot F \cdot C_{load} \cdot V_{dd}^2$$

$$P_{leakage} = V_{dd} \cdot I_{leakage}$$

TR = Toggle Rate $\quad$ F = Frequency

C_{load} = Capacitance $\quad$ V_{dd} = Voltage

V_{dd} = Voltage

$I_{leakage}$ = Leakage current (influenced by bias voltages)

All of the low-power optimizations that are discussed here work by reducing one of the above terms. Due to its quadratic relationship, the voltage is often the target of these optimizations. There are dozens of different techniques used to reduce the power consumption, but only a handful have an impact on the functional verification of the design. Table 3-1 lists the most common techniques and gives an indication of how they influence the power consumption of the device.

Table 3-1 Most Common Techniques for Low-Power Optimization

Name	Description	Leakage	Dynamic	Verification Impact
Clock Gating	Reduce switching activity by gating the clocks when logic is inactive		✓✓✓	low
Multiple Supply Voltages	Reduce the voltage level on slower logic	✓	✓✓✓	low
Power Shutoff	Dynamically turn off logic to portions of the design that are not needed.	✓✓✓	✓✓✓	high
Dynamic Voltage	Dynamically reduce voltage to save power. Frequency is often reduced at the same time	✓	✓✓	high
Standby Mode	Drastically reduce voltage, but keep high enough to maintain state	✓✓	✓✓	high
Active Bias	Bias the substrate to reduce power or improve performance	✓✓		low
Optimization	General category of synthesis, placement, and technology optimizations to reduce power. Includes LP cells, multi-voltage threshold, localized clock gating, etc.	✓✓	✓✓	very low

3.2.4 Low-Power Architectures

The following diagram depicts a small low-power design and is used to introduce the most common terminology and low-power architectures. This design has seven blocks at the top level: CPU, Memory, GPU, Ethernet block with TX and RX components at top level, and a control block that includes interrupt controllers and the power-management logic. All of this logic communicates through a global system bus.

Figure 3-3 Example Low-Power Design

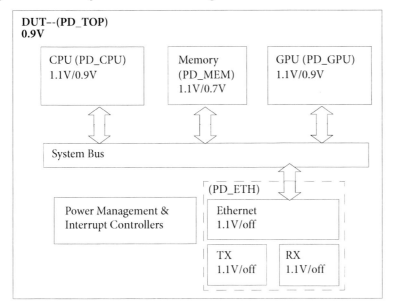

3.2.4.1 Power Domains

In the simplest terms, a power domain is collection of hierarchical instances that share the same power architecture. In the example above, the power domain names are listed in parenthesis. There are five different domains:

- PD_TOP—The default power domain for the design. All logic not placed into another power domain is automatically placed in this domain. In this case the domain is always on at 0.9V.

- PD_MEM—The memory is a power-shutoff domain. It can be turned on or placed in a low voltage sleep state dynamically.

- PD_CPU—This domain is a dynamic voltage domain. Depending on performance requirements, it can run at either 1.1V or 0.9V.

- PD_ETH—Highlights that a domain can contain multiple hierarchical instances. This domain is also a power-shutoff domain, but is controlled by a different shutoff condition than the PD_MEM.

- PD_GPU—A combination of both power shutoff and dynamic voltage.

Each of these types of domains is discussed in more detail later in this section. In a physical design, these power domains often become physical partitions and have their own power and ground network.

3.2.4.2 Power Modes

A power mode is one of a valid combination of the domains states and can be on/off or different voltages. For example, a valid mode in the above design is all powered on. Power modes have a great significance to verification. The more modes that are defined, the larger the verification space is for the design. It's very important to understand the power modes and the interactions between modes.

In the above design, the different domains are tightly related.

- The CPU and GPU are required to run at the same voltage.
- If the memory is in standby, then the GPU should also be off.
- The Ethernet controller is independent of the rest of the design.

The following table defines the different modes that would be valid for this design:

Mode Name	PD_TOP	PD_CPU	PD_GPU	PD_MEM	PD_ETH
NORMAL	0.9	0.9	0.9	1.1	1.1
ETH_OFF	0.9	0.9	0.9	1.1	OFF
HIGH_PERF	0.9	1.1	1.1	1.1	1.1
HIGH_PERF_2	0.9	1.1	1.1	1.1	OFF
POWER_SAVE	0.9	0.9	OFF	0.7(standby)	OFF
LOW_POWER	0.9	0.9	OFF	1.1	1.1
LOW_POWER_2	0.9	0.9	OFF	1.1	OFF

Each of these modes and the transitions between modes needs to be validated. The exact set of features and scenarios covered during each mode needs to be determined during a low-power planning session.

3.2.4.3 Power-Mode Transitions

Power-mode transitions are a way to specify which transitions between modes are legal. For instance, due to physical design reasons, it may not be possible to turn off the memory and the GPU at the same time. In this case, the transition from NORMAL to POWER_SAVE would first have to go through the LOW_POWER state.

3.2.4.4 Secondary Domain

One additional concept that is critical to modeling the low-power behavior is the secondary domain. There are special cells used in low-power design that derive their power from a secondary power source. These are cells like isolation and state retention for a power-shutoff design. The verification environment must model this power supply correctly or the functionality will not match implementation. The assignment of the secondary domain may not occur until late in the design process, so it is critical that the verification environment always uses the latest low-power intent.

3.2.4.5 Clock Gating

Clock gating is one of the oldest and most established methods to save dynamic power. Gating the clock obviously stops the majority of dynamic activity for the block in question.

Clock Gating Verification Considerations

For functional verification, the clock gating is part of the RTL and is simulated as normal. The only verification consideration is to ensure that the clock gating is enabled when expected and that all clock gating scenarios are covered during testing. Clock gating can essentially be treated as another mode or configuration of the hardware.

3.2.4.6 Multiple Supply Voltages (MSV)

Multiple supply voltages save dynamic and leakage power by reducing the voltage in sections that are not performance critical. Because the dynamic voltage has a quadratic relationship to the dynamic power, this method provides very significant savings. Because the two domains operate at different voltages, it is often necessary to insert a level shifter to match the voltages across the boundary.

Figure 3-4 Example Multiple Supply voltages Domain

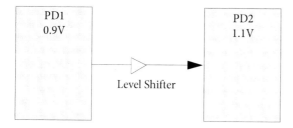

MSV Verification Challenge

While MSV has a number of physical design and verification considerations, from a functional verification point of view, it has very little impact. Static verification needs to verify that level shifters are specified if required on the interfaces between domains. The low-power design checks also ensure the proper power and ground connections.

3.2.4.7 Power Shutoff

Power shutoff is becoming one of the most common low-power techniques. As the name implies, power shutoff is a technique where power is physically shut off to a portion of the design. There is an actual physical power switch used, and when enabled, both the dynamic and leakage power effectively go to zero. This technique is common in many applications where a block of IP is idle and can be turned off when not in use. For instance, if a design has Bluetooth, the Bluetooth logic can be powered off when not in use.

Power shutoff is significantly more complex than MSV to implement and verify. Power shutoff includes three concepts: power switch, isolation, and state retention. Power shutoff comprises the following items:

- **Power switch**: A power switch turns on and off power to a group of cells, based on a control signal (shutoff_condition). A domain will have a network of physical power switches in the actual implementation; all of these are enabled by the same power control signal (or delayed versions of it). There is typically a significant delay between when the shutoff condition is removed and the domain actually reaches a functional voltage. For large domains, this delay is often measured in microseconds. In verification, the power down is modeled by corrupting the logic to "X" during shutoff and keeping it "X" until the power is fully restored.

- **Isolation**: Since the power is turned off in the domain, the logic downstream will essentially be undriven. This can cause an unpredictable value on the inputs and can also cause electrical shorts to occur. Isolation cells are used to protect against this and to ensure a known good value on the downstream logic. An isolation cell is powered by a different domain than the driving logic, and clamps the output to a specified value.

- **State retention**: State retention is a method for retaining some or all of the state elements during power shutoff. In some designs, this is accomplished by writing these registers into a non-volatile memory and then restoring that data on power up. For the purpose of this discussion, the technique discussed is a hardware solution called *state retention power gating*. This is technique uses a special type of flip-flop that contains a small retention latch that is loaded prior to power shutoff. The retention latch maintains its value during shutoff and is restored back into the main register after power is fully restored. There are several common scenarios for state retention:

    - Retain a small number of configuration registers. This is the most common scenario and allows the configuration to be maintained through the power cycle and speeds up the overall power-on process by not requiring the software to rewrite this configuration information.

    - Save all of the state elements. This second scenario is similar to a computer's hibernate mode, where the applications resume exactly where they left off before the power cycle. This is less common because it is expensive in hardware, and the small retention latch has leakage power during shutoff. On the hardware side, one of the factors to consider is the routing cost for both the secondary power supply and the save/restore control pins.

    - Do a "partial" state retention. In partial retention the designer attempts to save the minimum set of registers required to restore the design. This is not a recommended methodology because there is no way to formally verify that the set of registers was actually enough to restore the state of the design in all cases. The advantage, of course, is a reduction in the area/power/routing cost.

The diagram below shows the basic power-shutoff design. There is a separate power controller that is not part of the power-shutoff domain. The controller is designed to ensure isolation, and state retention is performed prior to powering off the domain. Details of the power controller and its verification are provided later in the chapter.

Figure 3-5 Example Power-Shutoff Domain

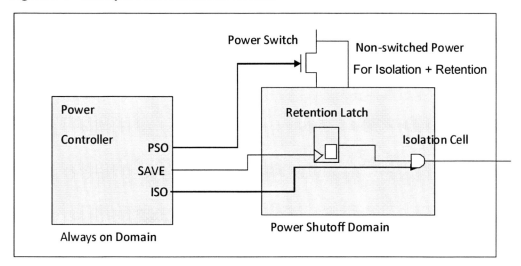

Power-Shutoff Verification Considerations

The following is a list of power-shutoff items to be considered in the verification task:

- **Increase in number of modes**—The design must work with this block on or off, and very often combinations of this block and others being on and off.

- **Modeling of power-shutoff corruption**—The verification engines should model the power shutoff of the domain and the impact of the secondary domains power state as well. This is typically done by forcing an "X" on the logic in the domain. This is a more complicated topic than one might expect, and is explored in more detail in the simulation section.

- **Modeling isolation and retention logic**—For power-shutoff conditions.

- **Verification of the power controller**—There is a required sequence for the control signals that needs to be checked. For example, a power-shutoff sequence is: clock gate–isolate–save state–power shutoff.

- **Valid isolation**—Ensure that all outputs were properly isolated. This includes making sure the output is isolated whenever the driving domain is off, as well as ensuring that it is isolated to the proper value.

- **Validate power up**—Ensure that the block exits the powered off state correctly, especially if state retention is used.

- **Validate any system-level sequences**—For example, does the block need to be clock gated before shutoff? Was it in an active transaction when shutoff occurred?

3.2.4.8 Dynamic Voltage

Dynamic voltage involves reducing the voltage to a domain based on the system demands or performance requirements. A simple example is a laptop computer that runs at full speed and voltage while plugged in, but runs at a reduced clock and voltage when on battery. Dynamic voltage is often coupled with a frequency change as well. Combined, this is called dynamic voltage and frequency scaling (DVFS).

Because DVFS targets the voltage and frequency of the design, it improves both the dynamic and leakage power of the design. The primary cost is in having two voltages and a means to switch between them.

Dynamic Voltage Verification Considerations

The following is a list of dynamic voltage items to be considered in the verification task:

- **Increase in number of modes**—The domain can take on multiple voltage levels each of these needs to be verified as well as any combinations of other domains. This can be quite substantial depending on the number of voltage levels allowed. For instance, if each domain can have 3 states, then the number of modes is 3^N instead of 2^N for power shutoff.

- **Transition in voltage needs to be carefully managed**—If the voltage is reduced before the clock is reduced, missed edges and bad results can occur. In many cases, the clock is gated during the transitions. In other cases, the user relies on the proper ordering of the transitions.

- **Timing and physical design**—This technique is particularly difficult for static timing analysis and physical design. The block has to be verified on the physical level in all of its possible voltage and frequency combinations.

- **Level shifting**—This may be required on the domain interfaces. The dynamic nature of the power transitions can make it difficult to determine if it is needed. Physical verification tools should be used for the final verification of these transitions.

3.2.4.9 Standby Mode

Standby mode is a special case of dynamic voltage. In this case, the domain is switched to the lowest voltage that will still maintain the state. But this voltage is so low that it cannot compute any new values and any change on an input will corrupt the logic. To prevent this, the clocks and inputs are typically gated.

Because the state is maintained, this technique does not require state retention and has very fast power-up times. The voltage reduction saves a large percentage of the static power, and the clock gating saves the dynamic power. Overall this technique is less effective at power savings than power shutoff, but is often is a good compromise where retention and quick power up are required.

Standby-Mode Verification Considerations

The following is a list of standby mode items to be considered in the verification task:

- All the issues associated with dynamic voltage apply to standby mode.
- Inputs are required to be stable while in the standby mode, the verification engines should automatically corrupt the logic based on activity.
- Level shifting may be required.

3.2.4.10 Active Bias

Bias is a design technique where a voltage level is applied to the substrate. There are many variations of applying bias, but for low power, it is typically done to reduce the leakage. Active bias is the ability to dynamically apply this bias.

Depending on the amount of bias and a number of other factors, the device may not be in a functional state when bias is applied. In these cases, the domain should be modeled like a standby mode, to indicate that the device will corrupt on activity. Very often, this technique is used in conjunction with the standby mode described above.

Active-Bias Verification Considerations

The active bias is similar to standby. It can effectively introduce new modes of operation that need to be verified, and may need to model the corruption based on activity. Outside of that, the verification challenges are more in the physical domain with timing, electrical connectivity, and power analysis.

3.2.5 Low-Power Resources

The discussion above is a very brief introduction to low-power design. It is in no way a complete description of all the low-power architectures and features that are in use today. There are a number of excellent resources, such as the *Common Power Format* and others available from the Si2 organization, that provide a more thorough education on low-power design:

- http://www.si2.org/www_site_map.php#LPC
- http://www.cadence.com/solutions/lp/pages/default.aspx

3.3 Low-Power Verification Methodology

The low-power verification methodology provides an approach that shows how to apply advanced verification methodologies to the verification of the low-power features and architectures. It is primarily focused on how to use a planning- and metric-driven methodology using the UVM to effectively verify a design. The methodology discussion also provides an introduction to the other components of the low-power flow, such as low-power design checks and equivalency checking, but it is not meant to be a comprehensive training on those topics. This methodology serves as a practical guide to setting up a verification environment for low power, what should be verified and how to do that verification.

Earlier, this chapter described the major areas of verification for low-power design. The next few sections elaborate on these areas, highlighting the verification challenges and the general approach to the verification.

Figure 3-6 Low-Power Verification Methodology

This section contains the following topics:

- **Low-Power Discovery**

    Understanding the low-power architecture and all the available features is the key first step to designing an efficient verification environment.

- **Creating a Verification Plan**

    Low-power design checks provide a formal, exhaustive method of verifying that the power intent is correct and complete. This section provides an overview of the checks available, and provides guidance on how to avoid duplication of effort between these checks and the UVM environment.

- **Creating a Power-Aware UVM Environment**

    Once a verification plan is created, how should one implement that plan in a UVM-based methodology? This section describes how to use the LP UVC to extend the environment into low power. It provides examples of common usage case for the UVC and explains the basic architecture.

- **Executing the Low-Power Plan in the UVM Environment**

    Once a plan of action has been established and the LP UVM environment is setup, the flow needs to execute to that plan. The execution involves running the various portions of the verification plan on the tools best suited to provide coverage. This can include LP checkers like linters, emulation, simulation, formal proofs, and so forth.

3.4 Low-Power Discovery and Verification Planning

This section discusses why an understanding the low-power architecture and all the available features is necessary to designing an efficient verification environment.

3.4.1 Low-Power Discovery

The first part of the planning process is to ensure that all interested parties have the same understanding of the architecture and the features available. The system architecture, physical design engineers, RTL designs, and verification team should meet and discuss the architecture, expected use cases, restrictions and requirements. This communication is important for any design, but for low power, it is even more critical.

There are often physical design constraints that determine the number of domains that can turn on/off at the same time, or specific timing requirements on power on/off that can result in design failures if not met. These details often evolve over time as the design is implemented, the communication between teams is critical to ensure that these constraints are represented in the verification environment.

The low-power discovery session typically starts with a block diagram of the design and the overall power architecture. For each IP and power domain in the design, the planning session should document all of the modes of operation, dependencies on other domains or IP, and expected behavior through the various power modes.

3.4.2 Verification Planning

Armed with this common understanding of the power architecture, a verification plan can be developed. There are many styles of verification planning and many ways to organize the low-power requirements. The following section provides a simple power-specific set of requirements and is designed to highlight the type of information that should be collected. The primary example is for a design with a reasonable amount of power modes, a case were each power mode can be defined in detail. In larger or more complex designs, this level of detail is typically provided for only a handful of the modes and a more hierarchical approach is needed.

Once a verification plan is decided upon, it needs to be integrated into the verification environment. The UVM LP extensions discussed later in this chapter facilitate the development of the low-power coverage, checking, and sequence generation.

3.4.3 System-Level Planning

At the system or major IP block level, the planning process is primary concerned with three things:

- Ensuring that the power management is operating correctly
- Ensuring that all power modes for the system have been executed and the behavior of the circuit validated in that mode
- Ensuring that all mode transitions have been exercised

3.4.3.1 Power-Mode Coverage and Domain Dependencies

At the highest level, the low-power architecture translates into a set of defined low-power modes. The verification plan needs to track that each mode and the transitions between each mode have been exercised. The power modes often come directly from the CPF definition and can either be mapped to a coverage model or that model can be automatically generated from the CPF definition. In more advanced cases, the generated coverage model can be used in conjunction with the other coverage models to define a project-level coverage model to verify the power modes in the context of the full-chip functionality.

In addition to covering the power modes, the verification environment needs to ensure that all the constraints on those modes and transitions are being enforced. A typical environment will employ additional metrics from assertions or other checkers to ensure this behavior.

3.4.3.2 Are There Dependencies Between Domains?

The low-power discovery should document any dependencies between domains. Those dependencies need to become part of the verification plan. Some examples are:

- Domain X must be off if Domain Y is off
- Domain Y's voltage must be greater than Domain Z

These types of checks help ensure the proper operation of the low-power management. If the power management attempts to place the domains in illegal states, these assertions/checks report the issue.

These types of dependencies can normally be inferred from the set of legal power states. But in more complex designs these simple properties are often easy to specify, facilitate the debug of the design, and serve as a way to qualify that the modes were indeed defined correctly.

Sequence Dependencies Related to Mode Transitions

Understanding the sequence requirements between power modes is critical for the correct electrical and functional execution of the design. For example:

- The GPU may need to power off before the Memory is put in sleep mode.
- The CPU and GPU cannot switch on simultaneously because they will cause a current spike.
- The Ethernet controller must be idle with no pending transactions before it can power off.

As the examples show, these requirements cover a full range of functional and physical requirements. The simultaneous switching of CPU and GPU is a physical design requirement that can be checked in simulation. The GPU is typically using the shared memory, so if the memory is off but GPU is on it may signal an error in the power management unit.

The planning process needs to define all of these cases as properties of the design. The verification team can build assertions or checkers to ensure that these properties hold true. Finally, coverage metrics will track how well the verification suite is exercising the logic.

The following figure shows a sample system-level power verification plan.

Figure 3-7 System-Level Power Plan

---

1 System Level Power Plan

1.1 Power Mode Coverage

Coverage: A coverage item that tracks all the low power modes.

1.2 Power Mode Transition Coverage

Coverage: A coverage item that tracks all possible transitions between domains

1.3 Power Domain Dependencies

Requirements between power domains. For instance, if the CPU is off the memory should be in standby mode.

1.4 Power Domain transition requirements

Requirements on ordering of power domain transitions or on simultaneous transitions.
Example: PD1 must complete its transition before PD2 can start its transition.

3.4.3.3 Power Management Planning

In a typical low-power system, there are several methods used to control the power state of a design. The interaction between these methods is a common source of error in the system. The verification environment needs to ensure that each of these methods is correctly exercised, including:.

- **Software control**—In many designs, the power manager is entirely controlled by software. Typically, the power manager has a set of control registers that are written by software, and these registers will trigger changes in the low-power state. Understanding and fully testing these registers is a basic requirement that needs to be captured in the plan

- **Localized hardware control**—Some IP blocks will have local control of power. For instance, it is fairly common for a peripheral to power off if the physical media is disconnected and, in some cases, if a simple timeout expires. When the device senses the disconnect, it does not need the software to put it into a sleep state.

- **Hardware process monitors**—In some advanced cases of dynamic voltage (called adaptive voltage scaling), the design incorporates analog hardware process monitors that track the current draw, temperature, and/or actual voltage seen in the device. Based on this information, the hardware can change the voltage and clock frequency supplied to a domain without any interaction with software or other digital hardware.

- **Hardware control**—A hardware module may include finite state machines (FSM) to control its low-power state. Instead of software controller the state, an IP's internal logic is used for control. An example is an asynchronous FIFO that has multiple voltage/frequency options. As the FIFO fills up, it can switch to a higher bandwidth (higher voltage and frequency) to keep up with demand.

The verification plan needs to identify all the methods of changing the system power mode and the state of each domain. The transitions between the power modes are typically where verification issues are detected.

Effective Utilization of the Low-Power Features

The low-power features have very real costs in terms of hardware and verification effort. The verification environment should ensure that the hardware is being effectively utilized. If a specific mode is not executed, or hardly executed, then the design and application have to be investigated. The cause may be a software error that prevents the hardware from doing specific mode transitions. Documenting the expected usage cases and mode profile in the verification plan allows the verification engineers to develop checkers that can catch these issues.

Typically, this level of analysis requires actual applications running on the system. The use of low-power emulation is highly recommended to provide the bandwidth to run applications on large designs. This analysis can be coupled with power-analysis flows to provide more detailed view of the power consumption and the effectiveness of the low-power techniques.

The following figure shows sample **power-mode selections of the verification plan**.

Figure 3-8 Power-Mode Selection of the Verification Plan

1.3 Power Mode 1 (replicate for each power mode)
1.3.1 Functional verification requirements

1.3.1.1 Scenarios/features to be tested in this mode
Provide a set of coverage items that track scenarios or features required to be verified in this mode. These coverage items are designed to validate that this mode of operation is functioning correctly.

   -Send a data packet to IP1

1.3.1.2 Properties that must be true in this mode
Assertions or other checks that ensure that this mode is correct. For instance, clock gating may be required for this mode.

1.3.2 Mode Transition Requirements

1.3.2.1 Entry requirements
Sequence checks that ensure a set of pre-conditions required before entering this mode. The intent is to have lower level checks that ensure that the power management hardware is operating correctly. Example: FSM is idle and no transactions are in the queue.

1.3.2.2 Transition kind
Coverage of events that can cause a transition from this mode. For example, a specific HW interrupt or

3.4.3.4 System-Level Power Behavior

The previous section discussed the verification of the low-power control and modes; this section explores the verification of the system while it is in each of these modes and during the transitions. It answers questions like:

- Can the CPU still run properly with the GPU turned off?
- What is the minimum requirement to prove that the design runs properly in mode X?
- Does a powered-off block present the proper state to the rest of the system?

Errors often occur when a powered-off block is isolated incorrectly and actively drives a system bus or presents a ready status to the outside world only to silently ignore any requests. There may even be cases where the low-power control logic itself gets powered off. These errors can lead to very unpredictable results on the domains they affect, and often prevent the domain from ever powering up again (without a full reset).

In most cases, this portion of the verification plan is more about defining the scenarios that must be tested, and less about adding specific assertions. Typically the non-low-power system-level verification environment will catch the issues noted above. Components like a protocol checkers, or scoreboards would detect the issues, assuming the scenarios were actually exercised.

Verification planning needs to define the use cases and what transactions and features need to be tested during each mode of operation. The emphasis here is on the interfaces and interactions between blocks. The internal operations of an IP or domain are generally well covered in other aspects of the plan. But whenever a feature employs logic from multiple domains, ensure that these are tested in the full context of the low-power solution. With that in mind, there is no need to verify every flavor of every transaction in every mode of operation. This would simply explode the verification space for the design without adding any additional coverage. The verification plan becomes the roadmap that defines the proper representative set of scenarios to ensure verification closure. The UVM LP extensions discussed later are used to help implement these low-power sequences and scenarios.

The verification plan should be hierarchical, and any IP that uses low-power techniques internally should be covered by a separate section in the plan. The ability to build a reusable verification for the IP block is one of the compelling reasons for applying the UVM to the low-power verification.

The overall low-power solution can also help here by formally proving that the interfaces have the proper isolation insertion. The low-power design checks can also ensure that the power control logic is correctly powered. Using these formal checks is more exhaustive and can improve debug and productivity.

3.4.4 Hierarchical Planning

The low-power intent is often hierarchical in nature. A block may have its own defined set of low-power modes and transitions. The verification plan should reflect this hierarchy. This makes the specification of the low-power modes simpler and allows the user to prioritize the testing more effectively.

3.4.4.1 Are There Domains/IP that are Independent from a Low-Power Perspective?

Low-power designs are faced with exponentially increasing complexity when all IPs and domains are related to each other. Identifying cases where the IPs do not interact allows the verification engineer to focus the verification effort on areas that will increase coverage.

In the example in Figure 3-3 on page 105 there were 4 dynamic domains: three with 2 states, one with 4 states. Without further analysis, this would translate to 24 modes (2x2x2x3). But in the example, the Ethernet controller only connects to the system bus; it has no connections between itself and other domains. Since the Ethernet doesn't interact, then it can be verified more independently. The number of modes to verify can be reduced by almost half and still have the same coverage. One could verify the permutations of the other three domains with the Ethernet on (12 modes) and then one additional test to verify the Ethernet off state. When one looks at the possible mode transitions, the improvements are even more drastic.

This situation occurs very frequently in multi-core designs. The cores connect to the same system components, but usually don't connect directly to each other.

The verification planning needs to define these relationships, and ensure that every design team agrees to them. If the verification team marks these as orthogonal and does not test them, but the actual hardware is connected, then design errors could escape detection.

3.4.5 Domain-Level Verification Planning

At the domain level, there are often additional requirements that need to be met to ensure proper low-power behavior. The verification plan needs to define this requirements and appropriate coverage metrics and checkers.

Power domain level verification planning consists of the following:

- Power control module requirements and sequence requirements translated into coverage and assertions.
- Internal state requirements for power transitions
  - The system level defined the top-level requirements for the system; these typically are external to a specific power domain. The requirements here are for the internals of this specific power domain.
  - Examples may be that the internal FSM be idle, internal buffers are empty, and so on. In many cases, the domain is designed with a clean power down flow, and going outside of this flow may cause errors or the inability to power up again. The designer of this block knows all of his assumptions, and can code these as assertions.
- Power-up requirements
  - Often a domain must be reset on power up, these power-up requirements should be documented in the plan and checkers put in place to ensure that behavior.
- State retention
  - Any state retention requirements need to be documented in the verification plan. In general, the CPF is used to specify and validate the retention. But the strategy should be described in the document

and agreed upon. The verification team should verify that the expected retention is in the CPF and define how this retention will be verified in simulation. This verification typically will not check each individual register, but rather check that the domain powers up and can process the next transaction properly.

- Isolation requirements
    - The CPF defines the isolation requirements for blocks. The verification plan should define what the general policy for the IP block. For specific control signals, the verification team may decide to explicitly check the isolation is working properly. Most other signals only have a requirement to be isolated; the actual value typically doesn't make a functional difference.

3.4.6 A Note on Verifying Low-Power Structures

The proper modeling of the low-power structures is an absolute requirement to the success of the low-power verification. It is tempting to try to use dynamic simulations to verify the low-level power structures like isolation and state retention. In general, this level of verification is primarily accomplished by way of the low-power design checks. While it is possible to create assertions for every isolated input/output and for every state retention element, this is generally viewed as unproductive. The verification tools read the power intent, and as long as that power intent is correct, the verification engines should model it correctly. In effect, this type of assertion is checking the verification engine and not the design. Checking the simulation engine's model of the low power design is important when adopting a new engine, as it is generally not run during the full regression suite because it can affect the simulation performance and does not provide any additional functional coverage.

Even at the gate-level, the preferred check is low-power equivalency checking. Equivalency checking exhaustively verifies that the implemented netlist matches the CPF specification

3.4.7 Recommendations for Designs with a Large Number of Power Modes

Today's designs may have thousands of unique power modes. It becomes impossible to test each of these modes exhaustively. A simple example is a multi-core design that has 16 cores, each with 3 modes of operation. This would translate result in over 4 million possible modes of operation. Clearly, a brute force approach to these modes is not feasible in emulation let alone RTL simulation.

3.4.7.1 Identify Independent Domains

The first line of defense for designs with a large number of modes is to partition the design into independent groups. In the example in Figure 3-3 on page 105, each power mode was independent of the rest of the power domains, and the number of modes to test was reduced by half. Applying this same strategy to a larger design can have orders of magnitude impacts on the number of modes.

3.4.7.2 Leverage Bus Architectures

A very common architecture is a set of IP blocks that only communicate through a shared system bus. Take the case where there are 8 IP's on the bus, and each IP can be powered on or off. This translates into 256 combinations.

However, because it is on a shared bus, often the verification environment simply has to start with all the domains on, and then iterate through, turning one domain on-off and off-on in series while a transaction occurs involving one of the other domains. This ensures that the domain in question does not lock up or interfere with other bus traffic while it is off or transitioning.

3.4.7.3 Prioritize the Power Modes

There still may be too many modes to verify exhaustively. In this case, the only resort is to prioritize the modes and ensure that the highest priority modes are targeted for validation. This prioritization needs be documented in the verification plan and needs to be agreed upon by all interested parties. When prioritizing the modes, one should consider:

- Common use cases—Cases that are likely to be used by applications running on the system
- Power savings—Target the modes that have the greatest power savings, which typically will expose more issues and ensure that the more aggressive techniques are properly functioning
- Market commitments—Committed items should be thoroughly tested
- Utility—The design team can help prioritize by providing feedback on whether specific modes make sense in an application or not. If the specific mode is unlikely to be utilized it should be lower priority.

3.4.7.4 Leaf-Level Verification

Verify that each IP can cycle through its power modes with the rest of the design in an ON state. This ensures that the basic low-power features for each domain work properly at the system level. For instance, it ensures that the isolation of a particular domain doesn't affect the simulation as a whole.

This is not a complete validation as it doesn't check to ensure that a combination of this domain with others being powered off does not affect the system.

3.4.7.5 Prioritize Interfaces Between Domains/IP

A large percentage of low-power issues occur at the interfaces between domains. In many cases, these are caused by missing isolation or isolation to an incorrect value. A good example of this is isolating bus control signals incorrectly to their active value. In many cases, this can lock up the system bus and often the entire simulation as well.

System buses in general are an example of where understanding the architecture can help with the verification planning. In module designs, the majority of the communication between IP blocks is through a shared system bus. If there were 8 power-shutoff IPs on the system bus, then one might be tempted to run all 256 different power combinations. But if the only communication between IP blocks is the system bus, then one can run a much reduced set. Sometimes it may be as few as the all-on, all-off, and iterations on each IP being off one at a time (10 cases). In some cases, you may want to verify a second domain powering off when one domain is already off. This would still drastically reduce the verification space.

Another advantage to this type of architecture is that the existing bus protocol checker will typically be sufficient to verify that the low-power behavior doesn't affect the bus.

3.5 Creating a Power-Aware UVM Environment

The previous section defined what needed to be verified. The next step is to define how to make a power-aware UVM environment that can implement that plan. To implement the plan, the environment needs to generate low-power sequences, integrate power into coverage metrics, and track the power mode of the system and the state of all the domains. At the same time, it needs to understand the CPF and rapidly adapt to changes in the low-power architecture. The methodology needs to fit into the overall low-power methodology and ensure that the verification environment's view of the power architecture is the same as what will ultimately be implemented.

Traditionally, users have accomplished this in an ad-hoc and often manual way. They create a verification model of the system by hand, define "modes" of operation that relate to the power state of the system, and manually try to track these modes. The verification environment, in effect, hard codes the power intent and must constantly be updated by hand whenever the power architecture changes. And, there is no way to be sure that the verification model matches the actual implementation.

The UVM LP methodology is designed to leverage the CPF file and generate a LP UVC (low-power universal verification component) that provides all of the hooks and data to accurately track the low-power state. It generates data structures and events that provide that information in the UVM. This allows the verification team to easily integrate low-power metrics and use events based on the low-power state in their verification environment. The LP UVC also provides a framework for driving low-power sequences into the DUT; as such, it provides a complete UVM-based low-power solution.

3.5.1 Tasks for a Low-Power Verification Environment

The tasks for implementing a low-power verification environment include:

- Ensure that the verification environment reflects the final hardware
- Track the low-power state of the design
- Add the low-power state to existing coverage items
- Use low-power events as triggers for checks in the verification environment
- Create verification sequences that wait for a low-power event before triggering the next part of a sequence
- Create sequences without detailed knowledge of the low-power-control signals (add a layer of abstraction from the low-power intent)
- Check that a sequence results in the proper low-power state
- Verify the hardware and software power management correctly transition the power states
- Create a reusable environment for IP

3.5.2 Solution: Low-Power UVC

The solution to low-power verification is to provide a low-power UVC (universal verification component) that is automatically configured based on the CPF. It contains a low-power monitor, basic low-power sequences, and a bus functional model to provide more advanced sequences for driving the power.

The monitor provides the UVM environment with access to the power state and events that trigger whenever the power state changes. Because the UVC is UVM based, it can be easily integrated into any UVM verification environment. For example, any functional coverage item can be crossed with the low-power mode from the UVC and ensure that a specific function was run in all power modes. The events on a power change can enable a user to check that configuration registers and state machines are in the proper state before or during low-power events. All this can be added without the verification engineer knowing all the low-level power-control signals or being an expert on low-power design.

The bus functional model provides a simple mechanism for a verification engineer to generate new low-power sequences. It provides a framework that includes simple transactions to switch to a specific mode. The user simply specifies "change to mode X", and the BFM (bus functional model) analyzes the current state, the mode transitions from the CPF, and can generate a sequence to effect that mode transition.

By default, the UVC has simple transitions that are based on the control signals defined in the CPF and can essentially force a specific low-power state. But this is not enough for a complete verification environment. The UVC is designed to be easily extended to define additional sequences to drive the power state. For most designs, this would typically include generating a bus transaction that writes to a power management register, and/or the generation of external requests or interrupts that result in a power change (example: a USB port that triggers an event when a device is connected or disconnected).

The following diagram shows how the LP UVC fits into an existing UVM-based verification environment.

Figure 3-9 Low-Power UVM Environment

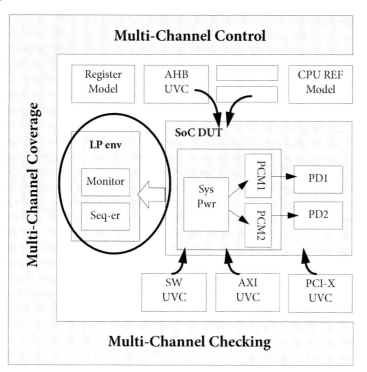

The LP UVC is generated based on the CPF and instantiated into the users existing UVM environment. It becomes the primary vehicle for both driving LP sequences as well as providing the rest of the environment with the low-power state of the design. In effect, it becomes an abstraction layer to the low-power data. As the low-power intent (CPF) changes, the generated UVC will reflect those changes and help ensure that the verification view of the power is always in sync with what will eventually be implemented.

The current implementation of the UVC is based on UVM *e* and will be extended to UVM SV in the future. The examples in this chapter are based on UVM *e*, but should be readily understood by SV users. The concepts and methodology surrounding the UVM LP UVC is independent of the choice of language.

Another way to ensure consistency in this flow is to use low-power aware system tasks available in a suitable simulator. In this case, the UVC does not track the low-power state itself, but instead, uses these low-power systems tasks to access the simulation engine's low-power state. Because the simulation engine calculates this state based on the CPF, the UVM environment matches the simulation environment, which matches the implementation. This greatly simplifies the UVC, and ensures a consistent level of support with the simulation engine. (If the simulator supports a CPF construct, the mode tracking in the UVC will also support it.)

3.5.3 UVC Monitor

The LP monitor is perhaps the most used feature in the LP UVC. It provides the access to the low-power state of the design and can be used in the generation of coverage, in LP checkers, and also to create complex sequences that depend on the power state.

The LP monitor is generated from the CPF and can be used out of the box—no modifications are needed to start taking advantage of the data it provides. Because it is generated by way of the CPF, it is always up to date. If a control signal or mode changes in the CPF, the LP UVC can be regenerated and have the correct data.

Before the UVM was developed, users would have had to create the logic that tracks the power mode of the system. This was a manual process that had to be updated frequently. Using the LP UVC provides this access using the UVM.

3.5.3.1 Using the Monitor for System-Level Checking

The monitor can be used to check the system-level behavior of the design. A simple example is a case where PD1 must be powered off before PD2 can power off. The monitor makes this type of check very easy to implement. It provides events that can be triggered on the power down of the domain, and also provides access to the status of the domain. The *e* code for this test is:

```
on lp_agent.PD1.power_down {
        check that lp_agent.PD2.status.state == SHUTOFF else dut_error(….);
};
```

3.5.3.2 Using the Monitor to Make Power-Aware Assertions and Scoreboards

A common question when discussing UVM LP is whether the existing scoreboards and protocol checkers need to be changed to support low power. The answer depends on how the checkers and scoreboard are set up. Most bus protocol checks should not have to change. Typically, the domain should not power off in the middle of a transaction, so any interruption of a transaction or bad status presented to the bus should be detected, regardless of power state.

For a scoreboard, it depends on the scoreboard and the design under test. If a functional unit is off, then the system may dictate that a new transaction should never be sent to it. In this case, the scoreboard may not need to be changed at all. In other cases, the expected result changes based on the on/off state. For instance, if unit is on, you expect the packet to go through successfully. But if it is off, then it should *not* get an acknowledgement that the packet was received and should error out if there was an acknowledgement. The UVM monitor can be used to provide the power state required to make the scoreboard power aware, and to distinguish between these cases.

3.5.3.3 What the Low-Power Monitor Is and What it Provides

The LP planning section discussed the need for power aware assertions, checking, coverage and sequence generation. The low-power monitor provides the data needed to generate these checks.

Power Domain Dependencies

Power domains often have dependencies between them. The low-power monitor can be used to easily check these dependencies. One example is that PDZ must always have voltage greater than PDX. This can be written in *e* as:

```
lp_agent.monitor. PDZ.status.voltage >= lp_agent.monitor. PDX.status.voltage
```

Using the Monitor to Create Coverage Using the Power Modes

The LP UVC contains LP coverage groups, which collect power related coverage information. Users can also utilize the power-mode information which is accessible through the UVC to define cross coverage connecting power status to data coming from other components in the verification environment like so:

```
cross lp_agent.monitor.lp_power_mode.cur_mode, some_functional_item;
```

Using the Monitor to Track Domain State and Trigger an Event Based on a Power Shutoff

The monitor also has events on the low-power state that can be accessed. These can be used to generate power-aware sequences. The following *e* example code checks the current voltage level, if it's at the low voltage, it sends a large packet. The system should detect the large packet and speed up the domain to process the package in a timely fashion.

```
if (lp_agent. PD1.status.voltage == 0.9 ) {
    do large_packet
    wait [1000]; // Expect the domain to speed up to process the large packet
    if (lp_agent.PD1.status.voltage != 1.1) {
        message ("Expected voltage to change");
    };
```

Low-Power Monitor Attributes and Events

The following table lists various low-power attributes and event information that are be provided by the monitor.

Level	Class	Name	Description
System	Events	lp_power_mode_changed	Top-level power mode changed
	Attribute	mode_name	Top-level power mode's current value
Domain	Events	lp_power_down	Triggered when domain is about to power off
		lp_power_up	Triggered when domain is powering up
		lp_power_standby	Triggered when the domain enters standby mode
		lp_mode_changed	Triggers when a power-mode group that this domain belongs to changes
		lp_nom_cond_changed	Triggers when the domain's nominal condition has changed
	Attributes	nominal_condition	Current nominal condition for the domain
		voltage	Current voltage for the domain
		state	Current sim state for the domain (on, off, standby, uninitialized)
		mode_name	Mode for the domain's power-mode group

3.5.4 LP Sequence Driver

The LP UVC includes a bus functional model (BFM) and sequence driver that are designed to provide an easy, customizable method to generate low-power sequences. The typical verification engineer interacts with this mode using very simple transactions.

Power-mode changes can be initiated in a number of ways in the typical system. The UVC automatically grabs control signals from the low-power intent (CPF), and can generate these base sequences automatically. In most cases, the verification team will generate a sequence library that reflects how the system is typically used. For example, most low-power designs employ software control of the power mode, and this is done by way of writing to a set of configuration registers. The user would simply create a low-power sequence that extends the automatically created one. When the library is built, the verification engineer can use the full breadth of the UVM to do pseudo-random generation of arbitrarily complex low-power sequences. One of the advantages is that this encourages reuse and only a few members of the verification team need to become low-power experts.

The UVC can also be extended to provide checks on each of the mode transitions. These checks can be used to ensure that any preconditions are satisfied before and after the transitions occur. For instance, the checker can check that the FSM is in an idle state before a domain powers off, or that clocks are gated. This ensures that the checks are done by anyone using the sequence library.

3.5.4.1 Using the Sequence Drivers

This section discusses using sequence drivers to change the power mode of the system, turning off and on individual domains, and toggling control signals.

Power-Mode Changes

The sequence driver can be used to change the power mode of the system. In the *e* example below, it uses a "VIA_BUS" mode that the user would have previously defined.

```
do VIA_BUS_CHANGE_MODE lp_seq keeping {
        .target_mode == PM_smc_uart;
      };
```

The UVM BFM analyzes the current power mode and finds a path to the `target_mode`. It uses the CPF definition of the mode transitions to determine a valid path, and executes one or more lower-level sequences to execute the transition. The verification engineer doesn't need to know that to get to the `ALL_ON` mode, the BFM actually had to transition to from `ALL_OFF` to `MODE Y` to `ALL_ON`. (For electrical reasons, it may not be possible to go from an `ALL_OFF` state to an `ALL_ON` state. The BFM derives this from the CPF and simply does the right thing).

Change Domain State

The sequence driver can also turn off and on individual domains. The following code segment turns off all of the power domains. This uses the power-control signals specified in the CPF to force a power transition. This is essentially a brute-force method that forces the power-control signals without regard for the current state or any other logic surrounding it. Typically, the verification team will extend these methods to provide a more careful power-shutoff cycle. But for early testing, this type of clever behavior can facilitate the verification.

```
do PCM_SHUTOFF_ALL lp_seq on driver.lp_sequence_driver;
```

Toggle Low-Power-Control Signals

The LP UVC also has access to the low-power-control signals defined in the UPF, and can be used to create LP sequences directly. It can send the isolation, save, shutoff sequence with varying delays to ensure that the design behaves properly. It can also send invalid sequences to ensure that the assertions and checking in the design are complete (negative testing).

3.5.5 UVM-Based Power-Aware Verification

The LP UVC provides the bridge between the advanced verification methodology offered by the UVM and the power-aware simulations enabled by low-power-intent languages like CPF. The LP UVC provides the abstraction layer necessary to enable the verification team to design with low power in mind. It allows the verification environment to observe the power state and to create more advanced low-power coverage metrics and checking. More complete and complex low-power scenarios that are influenced by this power awareness can be rapidly developed and deployed. The ability to drive the low-power state completes the verification picture. The UVM flow is enabled by this relatively small UVC. The final advancement is that this UVC is based on the common power intent and provides both automation and consistency with the full low-power design flow.

3.6 Executing the Low-Power Verification Environment

The previous sections defined how to define a verification plan that covers the low-power intent and architectures and how to use the LP UVC to implement the coverage and sequence generation to meet that plan. The final piece of the methodology is executing to that plan on the low-power verification engines.

The low-power verification methodology spans multiple tools and core engines. The one requirement that spans all of these engines is the need to model the silicon as closely as possible. The term "mimic silicon" is used to highlight the need to go beyond a simple model and mimic the actual silicon as accurately as possible. This is perhaps the greatest strength of the low-power verification methodology outlined here. The flow is based on the same golden power intent that defines what will be implemented and the holistic view of power across the entire design. The implementation flow enables the verification solution to accurately mimic silicon.

The full low-power verification solution is complex and employs multiple verification engines to optimize the verification tasks. The following sections discuss the various engines, their requirements and what they are designed to verify.

- LP design and equivalency checking
- Simulation and emulation technologies
- Automated coverage and assertions

3.6.1 LP Design and Equivalency Checking

The CPF file is as important an input specification as the RTL, and as such, quality checks on the CPF should be run prior to running any other verification or implementation tasks. The goal is to identify issues as early as possible in the design flow and to minimize the effort required to find and trace the problem to the root cause.

Even during early RTL development, it is important to make sure the power intent is specified correctly. For example, the ability to check the CPF specifications for correct isolation before simulation, can catch missing isolation cells before running a long simulation. The CPF checking should be the first step of any verification flow. These checks are formal checks and can identify problems as early as possible in the flow. This prevents costly design respins and reduces the overall verification time

Power-aware formal verification tools can be used to verify that the low-power architecture is complete and consistent and that the implementation of that architecture is correct. They can verify complex multiple million gate designs in less time than traditional simulation. Formal techniques do not rely on simulation vectors and hence, can provide more comprehensive checking of the design in less time than simulations.

Power-aware formal verification tools should be used throughout the design flow. The earlier a problem can be detected, the less expensive the resolution will be. Detecting problems early greatly reduces the time to market and the overall resources required for the solution.

3.6.2 Low-Power Structural and Functional Checks

The low-power checks are designed to identify as many issues as soon as possible. The checks cover a full range of low-power features, including features for clock gating, power shutoff (PSO), dynamic voltage and frequency scaling (DVFS), multiple supply voltages (MVS).

A few representative checks are listed below:

- Detect low-power-control signals that are improperly powered and will shut off when the domain shuts off. This can lead to unknown states of the power domain and will cause severe functional errors.
- Detect missing isolation logic between power-shutoff domains.
- Detect missing level shifters between power domains.
  The missing level shifter can cause logic values to not propagate properly between domains, causing data loss and functional issues.
- Detect missing or incomplete power connections in a physical netlist.
- Detect incorrectly implemented isolation logic, wrong power connection, different isolation level than specified in the CPF, or wrong placement of the isolation cell.
- Incorrect state retention cell implementation, wrong cell type, incorrect power connections, incorrect state retention control connectivity.

These checks are particularly important when any hand edits were done to the netlist, or third-party tools were used in the implementation of the design.

3.6.2.1 Low-Power Equivalency Checking

Low-power equivalency checking allows the RTL and CPF to be verified against the implemented design. This checking is impossible in a non-power-aware equivalency checker, because the RTL design does not contain the low-power logic. With power-aware equivalency checkers, the equivalency checking can be done through multiple iterations of the design and the multiple transformations the design undergoes in the implementation process.

The key to the success of this step is the ability to check multiple implementations of the same power architecture against each other. This allows the flow to verify the design even in the face of large changes in the implementation. For instance, implementation tools will often ungroup logical hierarchy to get the best optimization results. These types of design transformations need to be handled by the low-power flow.

3.6.2.2 Checking vs. Functional Verification

In many cases, there is overlap between what is provided by way of static checking and what can be done with dynamic simulation and assertion-based verification. The recommended methodology uses static quality checks whenever possible. The static checks are often very quick and exhaustive. Users will sometimes ask for automated assertions to check very low-level details, for example, check that a specific input is isolated to a specific value during shutoff or that the state retention latch correctly saves. But this type of check is using the simulator to check the simulator, and frankly, does not add significant coverage. (If the simulator interpreted the CPF incorrectly, the assertion would also be incorrect.) The flow does not recommend auto assertions in these cases, but while a user-generated assertion could add value, it is not required.

3.6.3 Requirements for Selecting a Simulator and Emulator

Functional simulation and emulation provide the heart of the functional verification solution. Dynamic simulation of the low-power logic is the only way to verify the complete behavior of the low-power design. This section outlines some of the requirements and considerations that must be met by the low-power engines, and provide some high level methodology suggestions for running those engines.

3.6.3.1 Accurate Modeling of the Low-Power Intent

The most critical part to the low-power simulation is accurately modeling (mimicking) the hardware. The CPF provides the power intent and the low-power solution ensures that what the simulation is modeling is what will eventually be implemented. This is a complex task that most simulators were not designed to handle. A partial list of what needs to be modeled for low-power follows:

- Tracking of power modes and domain states (voltage, corruption state, etc.)
- Power shutoff
  - Corruption of logic during power shutoff
  - Isolation modeling
    - Understanding the primary and secondary domain used for each isolation cell will be connected to in hardware (this affects when it will corrupt)
    - Understanding the role and function of each isolation cell
    - Modeling complex cases such as back to back isolation
  - State retention modeling
    - Modeling the save/restore and corruption of sequential elements
    - Advanced checking of the control signals
- Standby modes

- Corruption of the domain based on input activity
- Ability to enable/disable assertions based on power activity
    - Many assertions will fail when the domain they check is powering off; often these are false assertions that need to be enabled and disabled. (CPF provides a constraint to do this)
- Modeling corruption of enums, integers, reals, and wreal
- AMS simulations with power

3.6.3.2 All Simulation Runs Should Be Low-Power Enabled

If a design has low-power features, then all verification runs should be done with low-power modeling enabled. This is the only way to ensure that the power management is operating correctly and doesn't introduce errors. In some methodologies, low-power verification is treated almost an afterthought and only specific tests are run with low-power enabled. But this assumes that nothing in any other simulation run could possible cause a low-power related event. In practice, improper resets, writing the wrong bit of a configuration register, or any of a host of other issues could cause low-power events. If these happen in a non-LP simulation, then the odds are that the event would be ignored and errors would be not be detected.

This does require that the simulation engines see little or no performance degradation for low power. Otherwise the impact to the verification methodology is too great.

3.6.3.3 Holistic Flow

The low-power flow is a part of the full design verification and implementation flow. There is a risk that the implementation tools may implement the low-power logic differently than the verification tools. A holistic environment where the implementation and verification tools have been designed with a consistent understanding of the low-power intent is a tremendous benefit to the flow. The CPF- and UPF-based flows have been designed to help address this issue by using the same intent in all tools. The challenge is to ensure that each tool understands that intent in the same fashion.

3.6.3.4 Emulation in LP Methodology

Low power is typically a system-level issue, and often requires the actual application software to be running to accurately verify the design. The typical low-power simulation is done in a targeted fashion. This is effective at verifying that the design works in each mode, and that the mode transitions can occur. In reality, the power mode is usually determined by software and understanding the success of the power optimizations requires running real-world applications.

Low-power emulation takes the same low-power intent as used by simulation, and models that logic in the emulator. It provides all the performance of the HW emulation, and can quickly exercise the different low-power scenarios. It can also be used to provide more detailed power analysis when linked with power analysis or estimation technology.

3.6.4 Advanced Debug and Visualizations

The debug of low-power issues is a critical part of the low-power verification flow. It's not enough to simply model the low-power intent, but the user needs to be able to quickly evaluate the root cause of the problem. Requirements to consider are:

- Ability to determine when logic is driven by the virtual LP hardware in CPF or from the design logic. Isolation and power-shutoff corruption are two examples where it is important to understand the source.
- Ability to trace power-control signals back to the CPF and to the driving logic. In a CPF design, the power-control signals do not have to be present at each level of hierarchy. They are automatically connected from the source without specific ports. The debug features must allow the user to trace these control signals.
- Tracking power modes and domain states including voltage and corruption status
- Ability to see inside of state retention logic and track the "save latch"
- Capability to set break points on low-power events

3.6.4.1 Low-Power Browser and Waveform Display of Power State

To assist in understanding the overall power architecture, a power browser can be provided to show all the details about the power modes and power domains. Inside each power domain, details about what instances are covered, the isolation and retention rules, and the allowed state of each module can easily be displayed.

This low-power state is also important in the debug phase of the simulation. To debug low power, the design state needs to be included in standard waveform views. This allows the user to see the current state and to line up power events with regular signal activity. Figure 3-10 provides an overview of what SimVision provides for low power.

Figure 3-10 Overview of Low-Power Visualizations

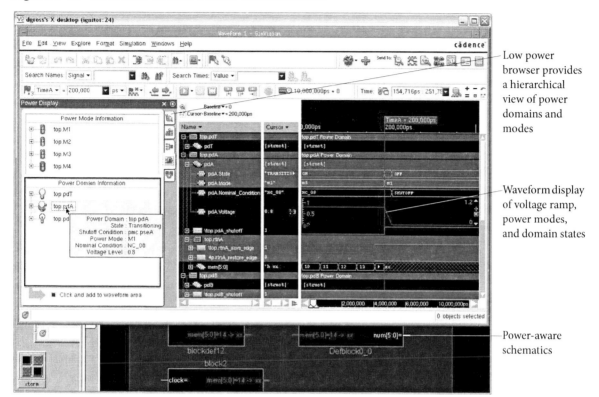

3.6.4.2 Visualizing Low Power's Influence on the Design

When low power is modeled in the design, it is extremely important that any signal tracing and visualization include the low power effects. For example, if there is an X in the simulation waveform, the user needs to quickly determine if that X is due to normal simulation or due to power-shutoff corruption.

Figure 3-11 Waveform of Power-Shutoff Corruption

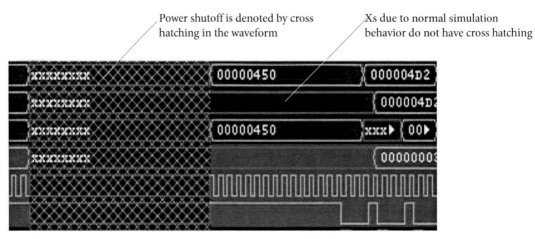

For isolation, it is equally important to understand whether a value on a net is due to that net being driven by an isolation rule or if it is normal behavior.

Figure 3-12 Waveform of Isolation

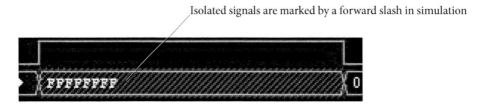

Another critical feature for low power debug is the ability to trace the contributors of a signal back to the source, whether that source is in the RTL or in the low-power intent. In Figure 3-13, a signal is currently X, due to power shutoff. But to debug why that signal is X, the user needs to understand why it is X. When SimVision traces the contributors of a signal, it includes the power-control signals that affect the logic. It clearly denotes the domain and control signals that are affecting the signals, and enables a very rapid debug of the logic.

Figure 3-13 Tracing a Signal Includes Low-Power Signals

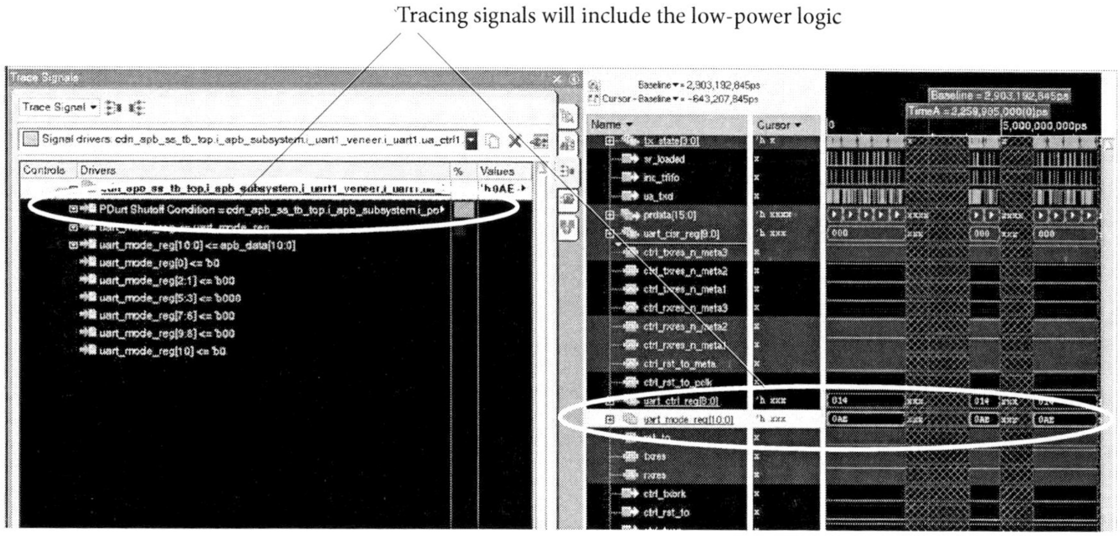

3.6.5 Automated Assertions and Coverage

The CPF can be used to generate a set of automated low-power checks and coverage. The language describes the power architecture and the low-level power-control signals. This information can be used by the simulator to generate a set of standard assertions and coverage items. These can be used to simplify the verification environment.

3.6.6 Legal Power Modes and Transitions

The CPF specification can define the set of legal power modes or configurations of a design, and can also define the set of legal transitions between power modes. Based on this definition, automatic checks can be generated to ensure that only the legal set is encountered during power simulations.

The ability to automatically generate these checks is critical. The power architecture will change over time as the design is optimized for low power. The automatic checking ensures that the checking is always in sync with the design specification.

3.6.7 Automatic Checking of Power Control Sequences

For power-shutoff design, a very specific sequence needs to be followed: first isolation, followed by state-retention, followed by power shutoff. On the power-up cycle, the exact opposite sequence needs to be followed, as shown in Figure 3-14 on page 136. This sequence is well defined by the CPF specification, and as such, assertions can be automatically generated.

Figure 3-14 Power-Up/Down Sequence

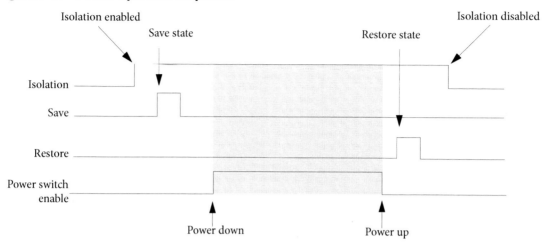

The checks listed below are a sampling of the checks that can be automatically generated by the CPF.

1. **Valid power-control signals**: The power-control signals should never be set to X's as this leads to undefined state of the entire domain and even affects the interfaces between domains. A simple assertion is to ensure that each defined control signal is never X.

2. **Isolation checks**: First, the isolation needs to happen before the power-shutoff event. Second, once a power domain is isolated, it should remain isolated during the entire power sequence. These checks can also be automatically derived from the CPF.

3. **State-retention checks**: If state retention is used, saving the state should happen prior to power shutoff, and it should never occur during a power shutoff (this would save X's). Similarly, the state should be restored only after the power is fully restored and stable. Restoring the state when power is off most likely indicates an error, as that restore will effectively be ignored.

4. **Complete sequence**: The power shutoff should include a complete sequence of isolation–save–power off–power on–restore–isolation off.

3.6.8 Verification Plan Generated from Power Intent

The verification plan completes the overall planning and metric-driven methodology. The plan is a simple standalone view of the verification progress during low power simulations. It provides an easy-to-read hierarchical view that has coverage of the system-level power modes defined in the power intent, and also domain-level coverage of all the power-control signals, domain states, and assertions that were modeled and checked based on the power intent.

Given the breadth of modern verification plans, many project teams are moving to automate the process. Doing so is best facilitated with a simulator that has the ability to generate a verification plan that includes all of the generated coverage and assertions described in the previous sections.

Verification Plan Generated from Power Intent

If both the verification plan and the coverage items are automatically generated by a verification planning tool, and based on the power-intent file, they are guaranteed be to be in sync with one another as the power-intent changes. This simplifies the job of the verification team, because the standard items for each domain are already tracked and well organized.

A tool-generated verification plan often provides features for the more advanced verification-planning users. It usually has a number of parameters that make it easy to integrate the plan into a user-defined verification plan. Typically, the low-power portion will just be one section of the larger system-level verification plan. To facilitate this integration, the plan is organized hierarchically, so individual domains and IP can be referred to in the appropriate section of the top-level plan. The parameters provide further refinement by allowing the user to turn on and off information that they care about. For instance, the user can selectively enable the power mode, domain, and assertion coverage, depending on how the top-level plan is organized.

Figure 3-15 shows an example of a tool-generated verification plan and the data it contains.

Figure 3-15   Automated Verification Plan

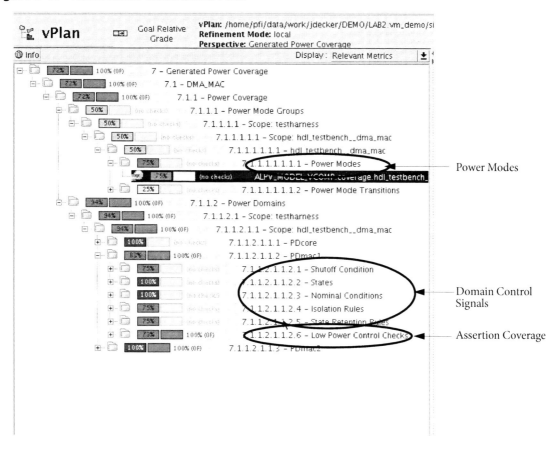

3.7 Common Low-Power Issues

The low-power architectures used in today's designs present many opportunities for failures. The following section provides a list of some of the most common issues found in designs. The verification methodology described here can be used to address all of them. The goal of this section is to provide some insight into the types of problems encountered so that verification engineers know what to look for and can perform the appropriate planning and checking.

3.7.1 Power-Control Issues

The low-power assertion section, "Low-Power Structural and Functional Checks" described the ability to generate assertions based on the low-power control logic. This section describes some of the common cases.

3.7.1.1 Oscillations in Power Shutoff Signal

In the figure shown below, the power shutoff is toggling throughout the power cycle. At a minimum, this would be a source of power consumption and noise in the system. It could also lead to electrical and functional issues caused by the unstable voltage. This error is detected by the automated assertions that track the power-shutoff sequence.

Figure 3-16   Oscillations on Control Signal for PSO

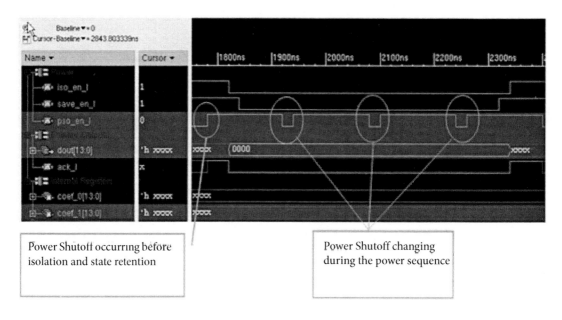

3.7.1.2 Premature State Restore

If state restoration occurs before the power is fully on, then the restoration does not take effect. An automatic assertion could also catch this case. In addition, the low-power modeling of the state retention would not restore the register and the Xs from the power-off corruption would remain in effect.

In this case, assertions provide a more intuitive check; they detect the root cause of the error. The X propagation provides a good functional model, but the user would need to interactively debug the design to get to the root cause of the Xs.

3.7.1.3 Power-Control Logic Corrupted

The power-control logic for one domain may actually belong to another power-shutoff region. If the power-shutoff controls for a domain are corrupted to X, what happens to the power switches? In hardware, if the input voltage to a power switch is active, then the power-shutoff control also needs to be active. If not, then the domain has a unknown state and can have a host of electrical issues. The low-power assertions also can detect this case and provide a clear message as to the cause of the failure.

3.7.2 Domain Interfaces

3.7.2.1 Incorrect Isolation

Isolation issues can cause a large number of design problems. Incorrect isolation values can present the wrong status to common buses, or can cause unexpected events downstream. These issues can be difficult to detect without a good low-power debug environment—that is, without debugging, it can be difficult to determine that an output is a '1' because of the logic driving it or because of an isolation insertion.

The following examples discuss a few cases of incorrect isolation that have been seen by verification engineers.

Example 1: Bus Hang Condition Caused by Incorrect Isolation Values

The isolation on the block is enabled, which causes the bus acknowledge to be asserted, as shown in Figure 3-17 on page 140. Because the acknowledge is actively asserted, no other transactions can take place on the bus. The bus controller eventually flags an illegal operation assertion because the bus protocol was not correctly followed. This is an example where an existing non-power aware UVC can detect issues caused by low power. The low-power visualization and debug help identify the root cause of the issue is the isolation.

Figure 3-17   Bus Hang Due To Incorrect Isolation

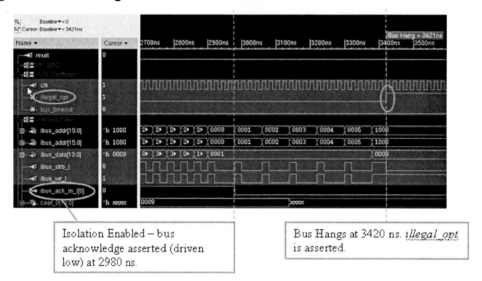

Isolation Enabled – bus acknowledge asserted (driven low) at 2980 ns.

Bus Hangs at 3420 ns. *illegal_opt* is asserted.

Example 2: Isolation Active at Wrong Times

Another case of incorrect isolation is when the power-control module incorrectly turns on isolation during normal operation. The output gets held to a fixed value and can lead to very unpredictable behavior. The low-power assertions can also help detect this issue.

Example 3: Data Loss

Incorrect isolation values can also cause data to be lost. Depending on the handshaking involved, the isolated value may signal that the block is ready to receive data, and packets of data could be sent to it. But if the domain is actually off, this data will not be processed. Of course, in most architectures, the handshaking on the bus would prevent these cases.

3.7.3   System-Level Control

The power-management control itself can also be a source of errors. The section on low-power planning mentioned several properties of the system that need to be verified. This section discusses some of the errors that may be encountered.

3.7.3.1   Improper Sequences For a Power Transition

There are two cases of improper low-power sequences:

- The system may attempt to turn on clocks or remove a reset before the power domain is fully up.
- The block has a pending transaction that isn't cleared before power off, and the transaction is lost.

3.7.3.2 System Fails to Take Proper LP Action

The system software may not exercise a low-power feature, either because the software is incorrect or the hardware does not send the proper request. In many cases, this is a failed opportunity to save power. In other cases, it may prevent the system from coming out of a lower power state.

3.7.3.3 Low-Power Domain Sequences

The order of domains powering up or domains powering up simultaneously are also causes of error in the design. These requirements need to be captured in the verification plan and assertions/checks can be developed to ensure the proper operation.

3.8 Summary

The use of advanced low-power design techniques creates a complex environment that needs to be verified throughout the design and implementation flow. Even when the low-power architecture is relatively simple, the number of power modes and the control of these modes can be quite complex. While accurate simulation is at the heart of the solution, it's not enough. The low-power verification methodology needs to take a more holistic approach utilizing all the available verification tools and techniques

The methodology needs to start with a comprehensive verification plan. Doing so allows the project team to comprehend both the hardware and software conditions that will drive power-mode changes and measure how completely they've been verified. This plan needs to go beyond traditional verification planning and include verification outside of the normal simulation based activities. The use of low-power design checks is required to ensure the proper low-power intent before simulation should even begin. For complex designs, the verification plan has to communicate the priorities and interactions between domains and power modes, and guide the verification effort to the most efficient solution.

From there, an accurate simulator is needed—but not only one that accurately reads the appropriate power format file, but one the mimics the silicon functionality. And it has to do that while running simulation fast enough to assure every verification test can be run power-aware. This last requirement cannot be stressed enough; the design is now RTL plus the low-power intent. To verify without the power intent is like verifying without all the RTL—the results are meaningless. Verification with the complete low-power intent, and with assertions and checks based on that intent, is the only way to ensure verification closure.

While the verification suite could be written as directed tests, doing so is unlikely to achieve the functional coverage necessary to have confidence in the end-product. A modern verification methodology like the UVM is required. One of the challenges for these low-power designs is to make that higher-level verification environment fully power aware and to ensure that it accurately reflects the architectures represented in the low-power-intent files. The UVM environment described above guarantees this consistency by reading the power intent and configuring its monitoring and checks directly from this power intent.

Finally, the solution needs to scale to the target environment including analog-mixed signal and system validation. The system validation often requires running actual software applications on the design. The use of power aware emulation technology is required to provide any type of reasonable coverage of these applications.

The verification of advanced low-power design is challenging and complex. Careful planning, accurate modeling and utilization of all the available verification techniques provides the only methodology for verification success.

4 Multi-Language UVM

This chapter describes the Universal Verification Methodology (UVM) Multi-Language implementation. The UVM Multi-Language (UVM-ML) defines how to create reusable verification components while working in environments using more than one language. These reusable verification components are called UVM Verification Components (UVCs).

To find out ...	Read ...
About UVM, UVC requirements, and using *e* and SystemVerilog in the same environment	"Overview of UVM Multi-Language" on page 143, "UVC Requirements" on page 146, "Fundamentals of Connecting e and SystemVerilog" on page 147, and "Configuring Messaging" on page 154
How to create an *e* layer on top of an existing SV Class-Based UVC	"e Over Class-Based SystemVerilog" on page 154
How to create multi-language environments, in which the low level of the UVC is an *e* UVC (*e*-implemented UVC) and the higher level is implemented in SystemVerilog Class-Based (UVM-SV)	"SystemVerilog Class-Based over e" on page 171
How to use the UVM SystemC library and its new features	"UVM SystemC Methodology in Multi-Language Environments" on page 174

Note UVM was designed for vendor-independent, multi-language operation.

4.1 Overview of UVM Multi-Language

UVM-ML uses UVCs working in an environment of more than one language. The environment for a UVC has three conceptual layers, which are described in Table 4-1.

Table 4-1 UVC Layers

Layer	Task	Units / Implementation	Description
3	Central control of multiple subsystems	• Multi-channel sequence • Scoreboard	Coordinates stimulus and checking of multiple subsystems
2	Generate test stimuli, perform protocol and data checks, collect coverage, and provide debugging information	• Sequencer (SV) Sequence driver (*e*) • Monitor	The sequencer (or sequence driver) generates stimuli and sends it to the driver (BFM). This upper level of the monitor performs checks based on information passed from the lower layer (layer 1).
1	Connect to the DUT, implementing the DUT interface protocol	• Driver (SV) BFM (*e*) • Monitor	The driver (BFM) gets input from the upper layers (2 and 3) and injects it into the DUT, based on the protocol. The monitor reads data from the DUT interface and passes collected information and events to layer 2.

In a UVC, each layer can be implemented in any language. Therefore, each layer should implement an interface to other languages. Figure 4-1 describes the UVC architecture. As shown in the figure:

1. Level 1 can be implemented in any language. Typically it includes a:

    - Unit for driving the stimuli according to the protocol
    - Monitor for monitoring the activity

    In this example, level 1 has three modes of interfacing to the DUT:

    - Signal level (Sig)
    - Transaction-level model (TLM)
    - Acceleration (Accel)

2. Level 2 can be implemented in any language. Typically, it includes:

    - Data generation
    - Sequences mechanism
    - The configuration struct (Config)
    - The vPlan interface to Incisive Enterprise Manager
    - Environment output: error messages (Chk), coverage reports (Cov), and debugging utilities (Dbg)

3. Level 3 is the user interface. Regardless of the language in which level 2 is written, level 3 in any UVC should support as many languages and user interfaces as possible. For example, it should support:

- *e*
- SystemVerilog (SV)
- SystemC (SC)
- Verilog or VHDL (HDL)
- File—any file type defined and supported by the UVC

Figure 4-1 UVC Architecture

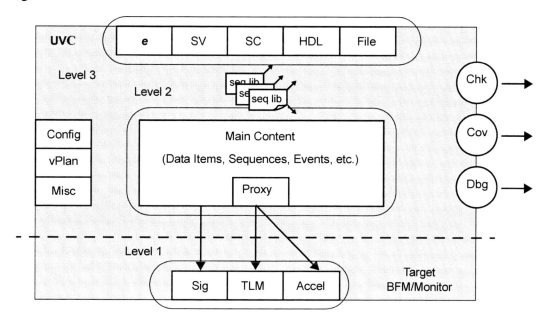

You can use several utilities to add a large part of the interface connections to your UVC, but you must manually create some of the layer connections when one of the interface UVCs is implemented in SystemVerilog. The following sections in this chapter provide detailed examples of UVCs and include:

- "e Over Class-Based SystemVerilog" on page 154 discusses creating a system *e*-UVC when one of the interface UVCs is implemented in SystemVerilog.
- "SystemVerilog Class-Based over e" on page 171 discusses the following topics:
  - Creating a SystemVerilog API to an existing *e*-UVC using the API Builder.
  - Creating a system SV-UVC when one of the interface UVCs is an *e*-UVC with a SystemVerilog API.
- "UVM SystemC Methodology in Multi-Language Environments" on page 174

For more examples and further information on multi-language verification, see the contributions area at: http://www.uvmworld.org.

4.2 UVC Requirements

This section describes the basic requirements necessary for a UVC to function in a multi-language environment. The details of implementing these requirements are described in the remaining sections of this chapter.

4.2.1 Providing an Appropriate Configuration

The UVC configuration should:

- Provide a central control over environment topology
- Provide an API for exporting and importing configuration information
- Provide an API for changing configuration at run time

4.2.2 Exporting Collected Information to Higher Levels

All UVCs should export important information collected during a simulation run. This information should be analyzed and used by other components in the verification environment. For example, the UVC might export information to scoreboard hookups. When working in a multi-language environment, the export of this information should be language independent, so that it can be read and used by any verification language. To achieve that, the UVC should:

- Export scoreboard hooks
- Send to scoreboard methods
- Export event notifications

4.2.3 Providing Support for Driving Sequences from Other Languages

Typically, a UVC has a sequence library. When a UVC is connected to a higher layer, the UVC must provide the means for the higher layer to activate the UVC sequence library. This eliminates the need to re-implement the sequences in another language, thus maximizing reusability and efficiency. To achieve this, the UVC should export its:

- Sequence driver (sequencer)
- Sequence item
- Sequence library (complete or a subset thereof)

4.2.4 Providing the Foundation for Debugging of All Components

When an environment is combined of several components, each implemented in a different language, debugging abilities become crucial and can significantly affect the efficiency of the verification process. The goal is to unite the debugging tools so that all components are debugged with the same tools. To achieve this goal the UVC should:

- Export a list of loggers
- Allow control of message logger behavior
- Allow filtering of messages based on various parameters
- Allow setting and clearing of trace commands
- Allow control of wave commands
- Provide central control of DUT error reports

4.2.5 Optional Interfaces and Capabilities

In addition to the requirements listed above, you can enhance the UVC quality and overall efficiency by providing interfaces to, or an implementation of, the following:

- Assertions for formal verification
- Acceleration
- Memory and registers package
- vPlan
- Reset management
- Test phase coordination
- Extended packaging

4.3 Fundamentals of Connecting *e* and SystemVerilog

Simulators provide the required means for connecting code implemented in two languages. This section includes discussions on:

- "Type Conversion" on page 147
- "Function Calls Across Languages" on page 150
- "Passing Events Across Languages" on page 153

4.3.1 Type Conversion

For every type defined in *e* that is transferred to or from another language (either an argument or a return value of a method that has a port bound to external), a corresponding type and default conversion in the other language should be created.

The SystemVerilog Adapter Configuration API lets users control the conversion of the exported types. Table 4-2 lists methodology recommendations for a typical adapter configuration.

Multi-Language UVM

Table 4-2 Recommended SystemVerilog Adapter Configuration

Items to be Configured	Method	Default	Configuration in Example UVC (vr_xbus)
Fields to be converted	pass_field()	All physical fields.	All fields
Types	get_type_name()	sn_ prefix added to *e* name Example: *kind* converted to *sn_kind*	Name is identical to name in *e*, with one change — a suffix of _s (standing for "struct") replaced with _c (standing for "class")
Enum value names	get_enum_item_name()	sn_ prefix and enum name added Example: *READ* converted to *sn_kind__READ*	Same as the default behavior, removing the sn_ prefix The type name prefix is necessary to avoid possible collisions in SystemVerilog
Field randomizability	randomize_field()	Converted field is not randomizable	All converted fields are randomizable
Access modifier	make_field_protected()	String and dynamic array fields are declared as protected	No field is protected
Parameter direction	get_parameter_direction_for_type()	Parameters to functions are passed as references	Pass structs as inout so that the interface functions can be called from automatic tasks

As an example, the following code specifies that:

- Names of the converted enum types are the same as in *e* (without the "sn_" prefix).
- All fields (not only physical fields) are exported.
- All fields are randomizable.

Code example:

```
extend sys {
    xbus_adapter_config : vr_xbus_config_unit is instance
    keep xbus_adapter_config.package_name == "vr_xbus";
};

unit vr_xbus_config_unit like sv_adapter_unit {

    // The name of the type in SV
    // Make enum type names same as those in e
    // For structs, replace the _s suffix with _c
```

```
        get_type_name(cur_type : rf_type) : string is {
            if(cur_type.get_package().get_name() ~ "vr_xbus") {
                result = cur_type.get_name();
            };
            if (cur_type.get_name() == "vr_xbus_api_trans_s" ) {
                result = "vr_xbus_api_trans_c";
            };
            if (cur_type.get_name() == "vr_xbus_api_resp_s" ) {
                result = "vr_xbus_api_resp_c";
            };
        };

        // The fields to be exported to SV
        // By default only physical fields are passed
        // Modify so that all fields are exported
        pass_field(cur_field: rf_field): pass_field_config_t is {
            var cur_struct: rf_struct = cur_field.get_declaring_struct();
            if (cur_struct.get_package().get_name() ~ "vr_xbus") {
                var struct_name := cur_struct.get_name();
                if (struct_name ~ "/.*vr_xbus_api*/") then {
                    result = YES;
                };
            };
        };

        // All fields of the data item must be randomizable
        randomize_field(cur_field: rf_field) : bool is {
            var cur_struct: rf_struct = cur_field.get_declaring_struct();
            if (cur_struct.get_package().get_name() ~ "vr_xbus") {
                var struct_name := cur_struct.get_name();
                if (struct_name ~ "/.*vr_xbus_api/") {
                    result = TRUE;
                };
            };
        };
    };
```

Assuming this is the data-item struct (containing two virtual fields and two physical fields):

```
    struct vr_xbus
    _api_trans_s {
        rand_addr : bool;
        %addr : vr_xbus_addr_t;
        rand_read_write : bool;
        %read_write : vr_xbus_read_write_t;
    };
```

The default conversion of this struct is conversion of the two physical fields:

```
class sn_vr_xbus_api_trans_s;
    sn_vr_xbus_addr_t addr;
    sn_vr_xbus_read_write_t read_write;
endclass
```

Based on the configuration shown above, the conversion of the struct is:

```
class vr_xbus_api_trans_s;
    rand bit rand_addr;
    rand vr_xbus_addr_t addr;
    rand bit rand_read_write;
    rand vr_xbus_read_write_t read_write;
endclass
```

4.3.2 Function Calls Across Languages

To call SystemVerilog tasks or functions from *e*:

- Instantiate **out method ports** in the *e* code.

To call *e* methods from SystemVerilog:

- Instantiate **in method ports** in the *e* code.

This chapter shows examples of multi-language UVCs. In particular, "e Supplying Config Information" on page 158 and "SV Pulling Configuration Information" on page 159 show the use of UVM method ports to call methods between *e* units and SystemVerilog component classes.

Following are some short and simple examples.

Example 4–1   Calling a SystemVerilog function from e

To call a SystemVerilog function from *e*:

1. Define the appropriate method_type in *e*.
2. Instantiate an **out method port** in a unit.
3. Constrain the hdl_path of the **out method port** to the function name.
4. Call the **method port**.

For example, assume you want to call the following SV function from *e*:

```
function automatic bit valid_address(int address);
    if (address > 100) return 1;
    return 0;
endfunction
```

The *e* code for calling this function is:

Function Calls Across Languages

```
// 1. Define the method_type.
method_type valid_address (address : int): bit;

extend cdn_sample_checker_u {
    // 2. Instantiate the port
    valid_address : out method_port of valid_address is instance;
        keep bind(valid_address, external);
        keep soft valid_address.agent() == "SV";
        // 3. Constrain the hdl_path
        keep valid_address.hdl_path() == "valid_address";

    check_status() is {
        // 4. Call the method port
        if valid_address$(current_address) == 0 {
           dut_error("Current address is illegal, " current_address);
        };
        // Address is valid - continue checking
        // ...
    };
};
```

Notes

- The TCM calling a task is a blocking call. You cannot run multiple tasks in parallel, because the port is blocked until the task ends. To support parallel calls, implement a synch mechanism in SystemVerilog, which consists of:

    - A function (called by *e* and bound to the method port).
      This function registers requests-for-task.

    - An always block.
      Whenever there is a new request-for-task, call the task.

  For a complete example, see "Invoking SV Sequences from e" on page 166.

Example 4–1   Calling a Method from SystemVerilog

For calling an *e* method from SystemVerilog do the following three steps in *e*:

1. Define the appropriate method_type in *e*.

2. Instantiate an **in method port** in a unit.

3. Implement the **in method port**.

Do the following step in SystemVerilog:

4. SV: Call specman.<*port hdl_path*>.

Advanced Verification Topics   **151**

For example, assume you want to get *e* configuration information from SystemVerilog code.

1. In *e*:

    a. Define a method_type taking a config struct argument (passed by reference).

    b. Instantiate an **in method port** of that method_type.

    c. Implement the method.

2. In SystemVerilog:

    a. Call the method port from the SystemVerilog **build()**.

    b. If the SV-UVC contains its own configuration class, translate from the configuration class passed from *e* to the original UVC configuration class.

    c. Configure and build the environment, based on the configuration information passed from *e*.

The following *e* code enables transfer of *e* configuration information:

```
// Define the method_type
method_type cdn_uart_get_config_mt(cfg: *shr_uart_env_config);

unit cdn_uart_mod_env_u like any_env {

    config: shr_uart_env_config;

    // Instantiate the method port
    get_config_port: in method_port of cdn_uart_get_config_mt is instance;
        keep soft get_config_port.agent() == "SV";
        keep soft get_config_port.hdl_path() ==
                            append("cdn_uart_get_config_", env_name);

    // Implement the method port
    get_config_port(cfg: *shr_uart_env_config) is {
        cfg = config;
    };
};
```

The following code calls the method `get_config_port()` from SystemVerilog:

```
class eocb_uart_sve extends uvm_env;

    cdn_uart_env uart0;

    // ... more instances and functions ..

    // configuration objects
    cdn_uart::shr_uart_env_config   e_uart_config;    // e version
    cdn_uart_cfg                    uart_csr;         // SV version

    virtual function void build_phase(uvm_phase phase);
```

```
        super.build_phase(phase);

        // Pull the configuration from the upper e layer
        // Call the method port
        specman.cdn_uart_get_config_UART0($sn_get_id(), e_uart_config);
        // Translate to UVC original config class
        cdn_uart_translate_e_config(uart_csr, e_uart_config);

        // Use config info passed from the e config
        set_config_int("*uart0.Tx","active_passive", uart_csr.tx_active);
        set_config_int("*uart0.Rx","active_passive", uart_csr.rx_active);

        // build the UART UVC
        uart0 = cdn_uart_env::type_id::create("uart0", this);
        // .. Continue building the environment
    endfunction // build
endclass
```

Notes

- For the input method port to return a value, define the argument as a pointer. In SystemVerilog, define the argument as `inout`. (For information on specifying argument direction, see "Recommended SystemVerilog Adapter Configuration" on page 148.)
- Sometimes, as in this example, the class SystemVerilog obtains from *e* is translated into another class. Typically, this occurs when SV-UVC code exists before adding the *e* code and therefore already has its own class.

4.3.3 Passing Events Across Languages

Use **event ports** for sensitivity to SystemVerilog events from *e* code.

To pass an event from SystemVerilog to *e*:

1. Instantiate an event port in a unit.

2. Constrain the hdl_path of the event port to be the SystemVerilog event that requires sensitivity.

3. Define an *e* event based on the event port.

4. Access the *e* event like any other event (using it as a sampling event for TCMs, wait/sync, and so on).

Example 4–1   *e* **Code Sensitive to SystemVerilog Events**

Your SystemVerilog environment contains an event, **address_phase**, that must be used in the *e* code.

```
    extend vr_xbus_bus_monitor_u {

        // Instantiate an event port
        sv_address_phase: in event_port is instance;
```

```
      // Connect to the required event using hdl_path
      keep soft sv_address_phase.hdl_path() == "address_phase";

      // Define e event, connected to the event port
       event normal_address_phase is
          {@sv_address_phase$} @sys.any;

      // Perform the check, using the event
      // Check that address phase detection comes only after getting transfer
      // information. If there is event indication with no transfer,
      // do not continue the run.
      on normal_address_phase {
         assert transfer != NULL else
            error("address_phase_sp event was emitted by the SV monitor, ",
                  "but there had not been any transfer start indication. ",
                  "Check the SV and e monitors integration");
      };
   };
```

4.4 Configuring Messaging

You can now configure messaging where:

- The internals of VIPs are implemented in *e* and the user-visible layer is in SystemVerilog
- You are using Multi-Language UVM

To configure messaging, use `uvm_message` at the simulator prompt. This command handles both SystemVerilog and *e* messages.

The options to the uvm_message command let you

- Filter messages based on verbosity level
- Enable messages according to verbosity and tag

Refer to the *UVM SystemVerilog User Guide* for the specifics on the command syntax and options.

4.5 *e* Over Class-Based SystemVerilog

This section describes the recommended flow for reusing a Class-Based SystemVerilog UVC in an *e* UVC or testbench. The typical use model for such an environment is creating an *e* testbench (or module/system *e*-UVC) re-using one or more interface UVC(s) implemented in SystemVerilog.

When using a SV-UVC within an *e* testbench, the following requirements must be addressed:

- Ability to configure the SV-UVC from the *e* layer during the construction of the SV hierarchy.
- Run-time control over the stimulus generated by the SV-UVC, including item and sequence-level generation, using the regular sequence constructs in *e*.

- Ability to connect to the monitors in the SV-UVC to add additional checking and coverage in the *e* layer.

This section uses a sample environment – a UART module UVC – to illustrate how the above requirements can be met. This environment verifies a UART transceiver with an APB host interface, as illustrated in Figure 4-2 on page 155. The module UVC is implemented in *e*, using two interface UVCs:

- UART SV UVC
- APB *e* UVC

Figure 4-2   Sample Environment: UART Module UVC

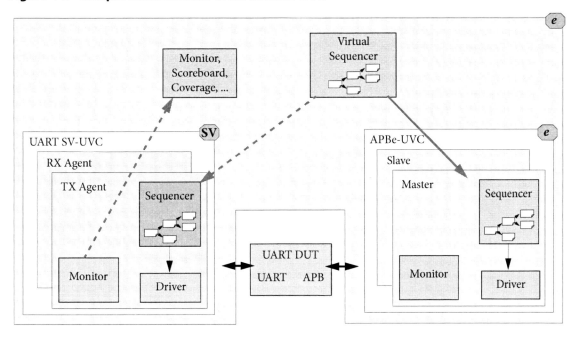

This section contains the following sub-sections:

- "Environment Architecture" on page 155
- "Configuration" on page 156
- "Generating and Injecting Stimuli" on page 160
- "Monitoring and Checking" on page 168

4.5.1   Environment Architecture

The architecture of a multi-language system (or module) verification environment is similar to the architecture of a single-language system verification environment. A system UVC is a reusable verification environment for a system. It contains module/subsystem UVCs, interface UVCs, and some additional verification components for the system level.

An interface UVC (for example, APB, PCI, or Ethernet) is a generic UVC for verifying the protocol on the interface.

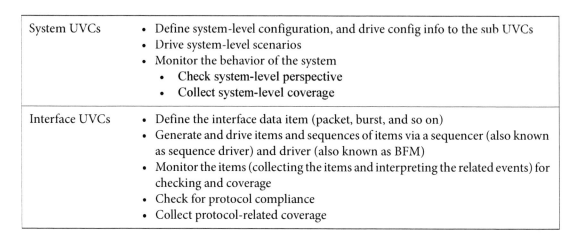

System UVCs	• Define system-level configuration, and drive config info to the sub UVCs
	• Drive system-level scenarios
	• Monitor the behavior of the system
	    • Check system-level perspective
	    • Collect system-level coverage

Interface UVCs	• Define the interface data item (packet, burst, and so on) • Generate and drive items and sequences of items via a sequencer (also known as sequence driver) and driver (also known as BFM) • Monitor the items (collecting the items and interpreting the related events) for checking and coverage • Check for protocol compliance • Collect protocol-related coverage

Figure 4-3 shows the architecture of a typical module/system UVC. It contains a monitor for getting information from the interface UVCs and a virtual sequencer for driving the interface UVCs' sequencers.

Figure 4-3 Typical Module/System UVC Architecture

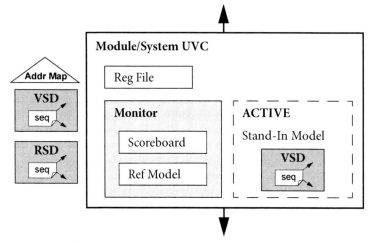

4.5.2 Configuration

Before the verification environment is operational, it needs to be generated (in *e*) or built (in SV). Configuration in this context refers to the ability of an integrator (or the test writer) to control various attributes in the verification environment that affect its topology and run-time behavior. It is recommended to propagate the configuration from top to bottom. Users configure the topmost component (system or module environment), and this component propagates the configuration information to the subcomponents. This methodology applies also for multi-language environments.

Although the term "top-to-bottom" is used, remember that there are several trees of components in the environment. There is the top SystemVerilog module, containing all SystemVerilog instantiating the DUT. There are then one or more top UVM components to instantiate the SV UVCs. Finally, there is also the *e* tree, starting from **sys**. Having multiple trees in different languages makes the order of build crucial to enable configurability.

To configure an SV UVC from the *e* layer:

1. Identify or declare the SV configuration class(es).
2. Export the configuration class(es) to *e*, and instantiate it so that it can be constrained during pre-run generation.
3. Pull the configuration information from the SV hierarchy during its build. The SV hierarchy must be built after the *e* environment has been generated.

4.5.2.1 Mapping SV Config Objects

It is recommended that you encapsulate all configuration information in one or more containers (structs in *e*, classes in SV). If the SV interface UVC does not contain a configuration class, you can create one.

The configuration class can then be mapped to an equivalent struct in *e*, either manually or using automatic utilities.

The following is an excerpt from one of the SV configuration classes controlling the UART SV-UVC.

```
// in package cdn_uart_pkg

typedef enum {baud9600, baud28800, baud57600 } baud_enum;

class cdn_uart_cfg extends uvm_object;
    randc baud_enum br_e;
    rand bit [15:0]   baud_rate_gen ;
    rand bit [7:0]    baud_rate_div ;
    rand bit [1:0]    char_length ;
    rand bit [1:0]    nbstop;
    ...
endclass
```

The following code snippet shows the mapped *e* configuration object, taken from
…/e/eocb_map_uart_cdn_uart_pkg.e

```
type baud_enum: [baud9600 = 0, baud28800 = 1, baud57600 = 2];

struct cdn_uart_cfg like any_struct {
    %br_e : baud_enum;
    %baud_rate_gen : uint(bits:16);
    %baud_rate_div : byte;
    %char_length : uint(bits:2);
    %nbstop : uint(bits:2);
    ...
};
```

4.5.2.2 Controlling the Build Order

The order in which the verification environment is built is critical to enable the passing of configuration values. The top *e* environment has to be generated first, to determine the desired attribute values in the configuration object(s). The resulting configuration object(s) can then be passed to the SV UVC(s), which must be built after the *e* environment.

You have full control over the order in which the *e* and SV portions of your environment are built. In this example, the **-uvmtest** switch is used to name the *e* test file, ensuring that the *e* environment is generated first. The top SV class is named with a **-uvmtop** switch, to make it generated second.

The following invocation command is taken from .../examples/uart_apb.irunargs, which is passed to irun as an arguments file (-f):

```
-uvm
-snsv
...
${PKG_DIR}/examples/testbench/tb_uart.sv
${PKG_DIR}/examples/config/eocb_sample_config.e

-uvmtest e:${eocb_test_file}
-uvmtop  SV:eocb_uart_tb
```

4.5.2.3 *e* Supplying Config Information

The *e* environment generates an instance of the required configuration object(s) mapped from SV. To make it available to the SV UVC, a UVM *in-method* port is defined. The SV logic invokes this port to pull the configuration object(s) during its build phase. The following is taken from the Incisive Verification Kit:

```
method_type cdn_uart_sv_uvc_cfg_mt(uart_cfg: *cdn_uart_cfg,
                         ve_cfg: *uart_ve_config);

unit eocb_env_u like any_env {
    ...
      config: cdn_uart_cfg;

    get_sv_uvc_cfg: in method_port of cdn_uart_sv_uvc_cfg_mt is instance;
        // enable the port as a UVM method port
        keep get_sv_uvc_cfg.external_uvm() == TRUE;
        //unique name for the utility object wrapping this port in SV
        keep get_sv_uvc_cfg.external_type() ==
                "eocb_env_get_sv_uvc_cfg_port";

    get_sv_uvc_cfg(uart_cfg: *cdn_uart_cfg, ve_cfg: *uart_ve_config) is {
        uart_cfg = me.config;
        ve_cfg = me.ve_config
    };
};
```

The above example illustrates the following guidelines:

- **Use one or more method ports to provide the mapped configuration object(s).**
  The configuration object (which was mapped from SystemVerilog) must be randomized during pre-run generation, and made available by the method port as either a return value, or arguments passed by reference.
- **Configure the method ports as UVM method ports.**
  This allows them to be called in a class-based fashion, unlike normal method ports.

4.5.2.4 SV Pulling Configuration Information

The SV logic that builds the environment must pull the configuration information from *e*. This logic is most naturally put in the testbench layer that instantiates the original SV-UVC and configures it. The SV testbench needs to use the UVM method port declared in the *e* layer, knowing the location of the port instance in the *e* hierarchy.

Note The configuration logic is non-reusable, and therefore resides under an examples directory in the UVC. In this multi-language example, the configuration logic is split between *e* and SV, where each portion is aware of the other's existence and location in the overall hierarchy. Note that the SV-UVC need not be modified, and has no configuration-specific code. There is nothing in the SV-UVC that is aware of the fact it is being used by an *e* testbench.

The following example shows the SV logic that pulls the configuration object and applies it to configure the SV-UVC. The code is taken from /examples/config/cb_tb.sv

```
class eocb_uart_tb extends uvm_env;
    string e_tb_path;     // e path to the counterpart e SVE unit

    // configuration objects
    uart_ve_config sv_ve_config;
    cdn_uart_cfg   uart_csr;

    // assigns path to the e SVE unit. Can be overriden in derived classes
    virtual function void set_e_tb_path();
        e_tb_path = "sys.eocb_tb";
    endfunction

    virtual function void build_phase(uvm_phase phase);
        // A UVM method port wrapper object declared by Specman
        specman_package::eocb_env_get_sv_uvc_cfg_port e_cfg_port;
        set_e_tb_path();
        super.build_phase(phase);

        // 1. Obtain the handle to the UVM method port, using its e path
        e_cfg_port = specman_package::eocb_env_get_sv_uvc_cfg_port::
                     get_port({e_tb_path, ".eocb_env.get_sv_uvc_cfg"});

        // 2. Invoke the method port
        e_cfg_port.get_sv_uvc_cfg(uart_csr, sv_ve_config);

        // 3. use the received configuration object(s)
```

```
        set_config_object("uart0", "ua_ve_config", sv_ve_config);

        // build the UART UVC
    uart0 = eocb_uart_env::type_id::create("uart0", this);
  endfunction
```

The above code illustrates the following guidelines:

- **The SV testbench (not the SV-UVC) handles the communication with the *e* testbench.**
  The UVC itself is left unchanged, and is unaware of the fact that it is controlled by another language layer.

- **The configuration object is pulled during build_phase().**
  The method port is invoked, and the received config attributes are applied (using `set_config_*`) during the build of the UVC.

- **The SV testbench must be aware of the *e* path to the configuration method port.**
  This is a manifestation of the fact that the testbench is split between two languages, with the configuration-specific logic present in both. Only the non-reusable, configuration-specific logic in the SV testbench needs to be aware of the *e* path, which is in-line with the UVM methodology guidelines. Note also that the *e* path is set within a virtual function (`set_e_tb_path()` above), which if needed can be overridden in classes that derive from `eocb_uart_tb`. In addition, the `set_e_tb_path()` function is called *before* the call to `super.build()`. This allows `set_config_string()` to be used as an alternative to assign a different value to the *e* path.

4.5.3 Generating and Injecting Stimuli

Stimuli generation and injection is done using sequences. The multi-language solution must create a user experience consistent with writing tests in a single-language environment — defining and **do**ing items and sequences.

The solution is based on layering an *e proxy sequencer* on top of the SV sequencer. The original SV sequencer in the SV-UVC need not be modified in any way. Instead, an *SV sequencer stub* (derived from a provided library base type) is instantiated, and controls the SV sequencer. The *e* proxy sequencer communicates with the SV sequencer stub directly, as illustrated in Figure 4-4.

Note There must be a one-to-one correspondence between instances of the *e* proxy sequencer, and instances of the SV stub. Not all sequencers in the SV-UVC must have a proxy in *e*. However, those that do must have exactly one instance of an *e* proxy sequencer and one instance of a SV sequencer stub.

Figure 4-4 Layering an *e* Proxy Sequencer on Top of the SV-UVC Sequencer

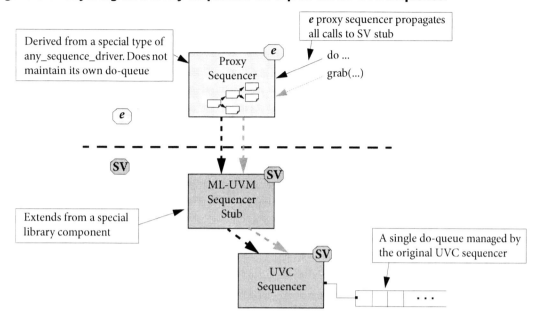

The example environment defines a virtual sequence driver connected to two interface sequence drivers. One is the sequence driver of the APB *e* UVC. The other is the proxy sequence driver connected to the UART SV-UVC. From the standpoint of the virtual sequence driver and its sequence library, the proxy sequences are exactly the same as ordinary *e* sequences.

4.5.3.1 Exporting a SV Sequence Item to *e*

To be able to reuse the SV-UVC sequencer, the relevant sequence item must first be mapped to *e*, similar to the configuration class earlier. However, mapping the original SV-UVC sequence item is not enough, since the process of randomizing it consists of several steps in a multi-language environment.

Randomizing a sequence item must involve constraints from both SV (within the UVC) as well as *e* (added by the integrator and test writer). However, currently, each language uses its own constraint solver, and therefore the randomization flow is split into two steps:

1. First, the sequence item is randomized in *e* with constraints on some (or all) of its fields. Each field, whose value is set in *e*, is marked to be used as-is by the SV-UVC. Other fields are left for the SV-UVC to randomize.

2. Next, the partially-randomized sequence item is passed to SV, where a second randomization step is performed. Fields marked to be used as-is from the first randomization step in *e* are left unchanged, while the rest are randomized based on the relevant SV constraints.

Note The above flow may lead to false contradictions, where some selections in the first randomization step, being done without access to the SV constraints, lead to a contradiction in the second step, even though the overall set of the *e* + SV constraints is not contradicting.

To implement the above flow, you need to add fields to the SV sequence item class, to mark the fields that were randomized in the *e*. It is recommended that you use a new field of type bit for each attribute that might be randomized by the *e* user. The name of the new field should be the original field with a use_ prefix. The UVC need not be modified. Instead, the new fields are declared in a new class that derives from the UVC sequence item, and is declared in a different SV package. If the UVC defines multiple sequence item types, you must repeat this for each of the UVC item types.

The following example shows the original sequence item definition in the SV-UVC. (The following code snippet is taken from the following location within the Cadence Incisive installation directory: kits/VerificationKit/soc_verification_lib/sv_cb_ex_lib/interface_uvc_lib/cdn_uart/sv/cdn_uart_frame.sv)

```
class cdn_uart_frame extends uvm_sequence_item;
    rand bit start_bit;
    rand bit [7:0] payload;
    bit parity;
    rand bit [1:0] stop_bits;
    rand bit [3:0] error_bits;
    ...
endclass
```

The following example shows the addition of the use_* fields to the sequence item. The code is taken from .../sv/cdn_uart_addons.svh

```
package cdn_uart_addons_pkg;
    ...
class eocb_frame extends cdn_uart_frame;
    bit use_start_bit;
    bit use_payload;
    bit use_stop_bits;
    bit use_error_bits;
    ...
endclass
```

The enhanced sequence item class (eocb_frame) is then mapped to *e* in the same manner as the configuration object. See "Mapping SV Config Objects" on page 157.

4.5.3.2 Creating an *e* Proxy Sequencer

With the sequence item struct available in *e*, you can now declare the proxy sequencer and connect it to the SV-UVC sequencer. This includes the following steps:

1. Declaring a proxy sequencer in *e*
2. Declaring a sequencer stub in SV.
3. Connecting the *e* proxy to the SV stub
4. Connecting the SV stub to the UVC sequencer.

Declaring a Proxy Sequencer in *e*

You declare the *e* proxy sequencer similarly to a pure-*e* one, using the sequence item struct that was mapped from SV. The only difference is the use of a special base type: `ml_proxy_seq_driver`. The following code example shows the declaration of a proxy sequencer, taken from .../e/eocb_uart_proxy_seq_driver.e:

```
sequence cdn_uart_sequence using
    item = eocb_frame,
    sequence_driver_type = ml_proxy_seq_driver,
    created_driver = cdn_uart_proxy_seqr_u;
```

Declaring a Sequencer Stub in SV

To avoid the need to change the SV sequencers, an SV stub component is used to facilitate the communication with the *e* layer. You must define a sequencer stub component type for each SV sequencer *type* that is to be controlled from the *e* layer. A special base type: `ml_uvm_seq::ml_uvm_sequencer_stub` is provided by Specman. The following example shows the declaration of a sequencer stub, taken from .../sv/cdn_uart_wrapper.svh:

```
package cdn_uart_wrapper_pkg;
   ...
class cdn_uart_seqer_stub extends ml_uvm_seq::ml_uvm_sequencer_stub;
   ...
   // Override this function to complete the randomization
   // of sequence items received from e
   virtual function void prepare_item_to_send(inout
                                       uvm_sequence_item item);
   cdn_uart_addons_pkg::eocb_frame  e_cdn_uart_frame;

      if (! $cast(e_cdn_uart_frame, item))
         `uvm_fatal("CAST_prepare_item_to_send", ...)

      // disable rand_mode for each field already randomized in e
      if (e_cdn_uart_frame.use_start_bit)
                 e_cdn_uart_frame.start_bit.rand_mode(0);
      if (e_cdn_uart_frame.use_payload)
                 e_cdn_uart_frame.payload.rand_mode(0);
      ...

      // randomize any remaining fields
      if (! e_cdn_uart_frame.randomize())
         `uvm_fatal("FAILED_RAND", ...)
   endfunction
endclass
```

The above code illustrates the following guidelines:

- **You must define a stub component per sequencer type.**
  Each sequencer type in the SV-UVC must have its own type of stub for it to be controlled from the *e* testbench. Multiple instances of the stub will then be created, according to the number of *e* proxy sequencers.

- **The most important functionality in the stub is its prepare_item_to_send() function.**
  This function is called to process a partially-randomized sequence item received from *e*, before it is passed to the SV-UVC sequencer. You must implement the logic within this function, and make it specific to the relevant sequence item.

- **The values set in the use_* fields control the randomization in SV.**
  The code within prepare_item_to_send() must examine the use_* field values passed from *e*. It should disable the randomization mode for all the fields for which the use_* counterpart field was set to 1. This disables the randomization of those fields, so that they preserve the value received from *e*. The last step is re-randomizing the entire object, to randomize a value for the remaining fields which were not controlled from *e*.

Accessing the ml_uvm_seq Package

The SV package ml_uvm_seq used in the previous section ships with the Specman installation. To use it, you must add the following command-line options to your compilation flow, as additional command-line arguments to irun or ncvlog:

```
setenv ML_SEQ_DIR `sn_root -home`/src

irun \
    ... \
    ${ML_SEQ_DIR}/ml_uvm_seq.sv \
    -incdir ${ML_SEQ_DIR}
```

Connecting the *e* Proxy Sequencer to the SV Stub

The *e* proxy sequencer is connected to the SV stub by assigning its external_uvm_path() attribute. This attribute is similar to hdl_path() in that it is concatenated with the values assigned in all the ancestor units. The string assigned to this attribute should name the UVM path to the SV stub in the SV components hierarchy. For example, in the UART workshop example, a sequencer stub is created in the SV hierarchy with the path: eocb_uart_tb.uart0.tx_seqer_stub.

Note In pure SV environments, component paths typically begin with uvm_test_top as they are instantiated under a root test component. However, in multi-language environments where a top component can be instantiated using -uvmtop, as in our example, the path does not start with the uvm_test_top instance.

For modularity, the relevant portions of the path can be assigned in the *e* code at various levels, as long as the concatenation creates the full string. The following code snippets illustrate this:

```
//in file .../examples/config/eocb_sample_config.e
extend sys {
    eocb_tb: eocb_tb_u is instance;
```

```
        keep eocb_tb.external_uvm_path() ==   "eocb_uart_tb";
    };

    unit eocb_tb_u {
        eocb_env : UART0 eocb_env_u is instance;
        keep eocb_env.external_uvm_path() ==   "uart0";
    };


    // in file e/eocb_env.e
    extend eocb_env_u {
        uart_tx_seqer: TX cdn_uart_proxy_seqr_u is instance;
        keep soft uart_tx_seqer.external_uvm_path() ==   "tx_seqer_stub";
```

In the above example, the values assigned to the `external_uvm_path()` attributes throughout the unit tree produce the full path to the SV stub (`eocb_uart_tb.uart0.tx_seqer_stub`) in the *e* proxy sequencer instance (`sys.eocb_tb.eocb_env.uart_tx_seqer`).

Connecting the SV Stub to the UVC Sequencer

The SV testbench must instantiate the appropriate number of SV stub components, one instance per each *e* proxy sequencer that will be connected to it. Each such stub must be connected to the corresponding sequencer in the UVC components hierarchy. The connection is performed by calling the stub's `assign_sequencer()` function during the connect phase.

The following example illustrates this. It is taken from /sv/cdn_uart_wrapper.svh:

```
    class eocb_uart_env extends cdn_uart_env;
        cdn_uart_seqer_stub tx_seqer_stub;

        virtual function void build_phase(uvm_phase phase);
            super.build_phase(phase);
            set_config_int ("Tx.sequencer", "count", 0);
            tx_seqer_stub = cdn_uart_seqer_stub::type_id::
                                    create("tx_seqer_stub", this);
        endfunction

        virtual function void connect_phase(uvm_phase phase);
            super.connect_phase(phase);
            tx_seqer_stub.assign_sequencer(Tx.sequencer);
        endfunction
```

The above code example illustrates the following guidelines:

- **Instantiate SV stub instances as required in the testbench around the UVC.**
  Each UVC sequencer that is to be controlled from the *e* layer by a proxy sequencer must have a corresponding stub instance.
- **[Optional] Configure the count attribute of the sequencer to 0, to disable its default activity.**
  Without this setting, the sequencer will start the default sequence (random) and generate random traffic which will be intermixed with transaction coming from the *e* layer.

- **Connect the stub instance to the actual UVC sequencer using assign_sequencer().**
  This provides the stub a handle to the UVC sequencer for it to control.

4.5.3.3 Doing SV Items from *e*

With the *e* proxy sequencer in place, doing an item on the UVC sequencer is very similar to doing an item on a pure *e* sequencer. The only difference is the need to constrain the use_* fields — to distinguish between attribute values that are generated in *e* from attributes that are to be further randomized by the SV-UVC.

The following example, taken from e/eocb_uart_proxy_seq_lib.e, shows a new sequence declaration for the *e* proxy sequencer. The new sequence includes one do action to execute a valid frame:

```
extend cdn_uart_sequence_kind: [ VALID_FRAME ];

extend VALID_FRAME cdn_uart_sequence {
    payload: byte;
    !frame: eocb_frame;

    body() @driver.clock is only {
        do frame keeping {
            .legal_frame == TRUE;
            .payload == me.payload;
            .use_payload == 1;
        };
        message(MEDIUM, "*** Frame Done ***") { print frame; };
    };
};
```

In the above example, the payload field originates from the SV sequence item and has a corresponding use_payload field. Because the sequence controls the payload value, it also constrains the use_payload to 1, signaling to the SV layer to use the payload value as is. All the other fields, for which the use_* fields is not set to 1, will be ignored in SV.

As a different example, the legal_frame field was added as an extension of the *e* struct (see /e/eocb_uart_sv_type_ext.e) and does not exist in the SV layer. Therefore, it is not passed down to SV, and has no corresponding use_* field.

4.5.3.4 Invoking SV Sequences from *e*

Most UVCs contain sequence libraries. To invoke or **do** these sequences from the *e* proxy sequencer you can export some or all of the sequences. Once a sequence is exported from SV, it can be invoked in *e* in the same manner as any ordinary *e* sequence.

To export a SV sequence, you must do the following:

1. Define an equivalent *e* sequence with the corresponding control fields. The sequence *must not* implement the body() method.

2. Extend the adapter config unit and add logic to associate the newly defined *e* sequence with its counterpart in the UVC.

The following code shows a SV sequence declared in the SV-UVC sequence library. (The following code is taken from the following location within the Cadence Incisive installation directory: kits/VerificationKit/soc_verification_lib/sv_cb_ex_lib/interface_uvc_lib/cdn_uart/sv/cdn_uart_seq_lib.sv)

```
class uart_incr_payload extends uvm_sequence #(cdn_uart_frame);
    rand int unsigned cnt;
    rand bit [7:0] start_payload;

    ... // constructor, object utilities, etc.

    virtual task body();
        ... // sequence logic here
    endtask
endclass
```

The following code taken from e/eocb_uart_proxy_seq_driver.e exports the above SV sequence, and makes it available to the *e* testbench:

```
extend cdn_uart_sequence_kind : [MAPPED_INCR_PAYLOAD];

extend MAPPED_INCR_PAYLOAD cdn_uart_sequence {
    cnt : uint;
    start_payload : byte;
};

// Associate the above with the actual exported SV sequence
extend eocb_adapter_config {
    post_generate() is also {
        associate_sequence("MAPPED_INCR_PAYLOAD
                            eocb_uart_apb::cdn_uart_sequence",
                           "cdn_uart_pkg::uart_incr_payload",
                           {"cnt"; "start_payload"});
    };
};
```

The above code illustrates the following guidelines:

- **Declare an exported sequence as a regular *e* sequence of the proxy sequencer.**
  The only key difference is *not* implementing the body() method.
- **Declare matching fields for all the required control fields in the SV sequence.**
  Every field that controls the exported SV sequence, or is to be read when the sequence ends, should be declared in the *e* sequence struct. In the above example, these are the cnt and start_payload fields.

  Note You do *not* have to map all the SV sequence fields to *e*. However, for those that you do, you must maintain the exact same field *name*, and use an equivalent *type*. The order of field declarations does not have to match that in SV.
- **Extend the adapter config unit to associate the *e* sequence to its SV counterpart.**
  The associate_sequence() method of the adapter config unit creates that association. It takes three

arguments: The *e* sequence type (fully qualified name is recommended), the SV sequence type (must use a fully qualified name) and the list of fields that will be passed to and from SV.

4.5.4 Monitoring and Checking

The monitor in a system or module UVC is a critical component in the verification environment. It obtains and maintains events and data related to the activity in the DUT. This information is made available to checkers, scoreboards, and coverage collectors. The monitor collects most of the information from the relevant DUT interfaces, or the monitors of underlying UVCs.

The connection of a monitor in the *e* layer to the corresponding interface UVC monitors is simply done using TLM ports. Interface SV-UVCs that follow UVM recommendations have monitors that put collected data items on TLM analysis ports. The *e* layer connects to those analysis ports directly, using interface ports in *e*.

To connect the *e* layer monitors to the SV-UVC monitors, you must do the following:

1. Identify the TLM ports in the SV-UVC which are of interest to the *e* testbench.
2. For each of those, define a matching interface port in the *e* testbench.
3. Register the selected TLM ports in SV with the ml_uvm library, so that they can be connected to an implementation in another language.
4. Connect the SV port with the *e* interface port.

4.5.4.1 Sample SystemVerilog TLM Port in the Monitor

As an example, consider the following analysis port defined in the UART SV-UVC monitor class. (The following code is taken from the following location within the Cadence Incisive installation directory: kits/VerificationKit/soc_verification_lib/sv_cb_ex_lib/interface_uvc_lib/cdn_uart/sv/cdn_uart_monitor.sv)

```
virtual class cdn_uart_monitor extends uvm_monitor;
    ...
    uvm_analysis_port #(cdn_uart_frame) frame_collected_port;

    function new (string name = "", uvm_component parent = null);
        ...
        frame_collected_port = new("frame_collected_port", this);
    endfunction

    task collect_frame();
        ...
        frame_collected_port.write(cur_frame);
    endtask : collect_frame
endclass
```

The above shows the monitor component declaring an analysis port (frame_collected_port). The `collect_frame()` task invokes the port by using its `write()` function to pass a frame whenever one is collected. Listeners (zero or more) connected to the analysis port will be notified whenever the `write()` function is called.

The SV-UVC in the UART example instantiates two such monitors – one for RX, another for TX. Therefore there will be two instances of the above analysis port which need to be connected from the *e* testbench.

4.5.4.2 Declaring a Matching Interface Port in *e*

You must declare an equivalent interface port in *e* to be able to connect to an SV port. This involves:

1. Mapping the SV object passed by the port to *e*. In most cases, this would be the same object type which you already mapped when exporting the sequence item, so no further mapping is required.

2. Declaring the matching interface port in *e*.

The following example shows the declaration of two interface ports in *e*, using the mapped sequence item type. This code is taken from…/e/eocb_mon.e:

```
unit eocb_mon_u {
    rx_frame_collected_ap: interface_imp of tlm_analysis of cdn_uart_frame
                            using prefix=rx_ is instance;
    keep bind(rx_frame_collected_ap, external);

    rx_write(frame: cdn_uart_frame) is {
        message(LOW, "[TLM] Notified of RX frame: ") { print frame; };
    };

    tx_frame_collected_ap: interface_imp of tlm_analysis of cdn_uart_frame
                            using prefix=tx_ is instance;
    keep bind(tx_frame_collected_ap, external);

    tx_write(frame: cdn_uart_frame) is {
        message(LOW, "[TLM] Notified of TX frame: ") { print frame; };
    };
```

The above code illustrates the following guidelines:

- **Use the relevant mapped *e* struct type.**
  In this case, the TLM port in SV is declared using the cdn_uart_frame class, and the *e* interface port matches.

- **Use the prefix option to support multiple port instances within the same unit.**
  The prefix allows different declarations of the `write()` callback, for multiple instances of the same TLM interface.

4.5.4.3 Registering the SV Port for External Connection

To enable the SV-UVC TLM port to be bound to an implementation in another language, the port must be registered with the ml_uvm library. This registration should be done in the SV testbench that instantiates the UVC, without the need to change the UVC itself.

The following code example is taken from …/examples/config/cb_tb.sv:

```
import ml_uvm::*;

class eocb_uart_tb extends uvm_env;
    function void connect_phase(uvm_phase phase);
        ml_uvm::external_if(uart0.Tx.monitor.frame_collected_port,
                        "cdn_uart_frame");
        ml_uvm::external_if(uart0.Rx.monitor.frame_collected_port,
                        "cdn_uart_frame");
    endfunction
```

The above example shows the SV testbench registering two TLM ports – one in each of two instances of the monitor shown in "Sample SystemVerilog TLM Port in the Monitor" on page 168. The `external_if()` function does not actually connect the port to anything, but only makes it possible to connect the port to an implementation in another language.

Note The ml_uvm package is automatically compiled by IES whenever you use one of these command line switches: -uvmtop, -uvmtest, -ml_uvm. You must also import the library at least in one place in your SV code.

4.5.4.4 Connecting the e and SV Ports

The last step is to connect the TLM port in the SV-UVC and the *e* interface ports. The ml_uvm library allows the connection itself to be done from any language. Because it is more natural to do so at the top layer which controls the configuration, the connection itself is done in *e*.

The following example shows the connection logic. It is taken from …/e/eocb_env.e:

```
extend eocb_env_u {
    monitor : eocb_mon_u is instance;

    connect_ports() is also {
        ml_uvm.connect_names(
          append(me.full_external_uvm_path(),
              ".Rx.monitor.frame_collected_port"),
          monitor.rx_frame_collected_ap.e_path());

        ml_uvm.connect_names(
          append(me.full_external_uvm_path(),
              ".Tx.monitor.frame_collected_port"),
          monitor.tx_frame_collected_ap.e_path()); };
};
```

The example above illustrates the following guidelines:

- **Use the connect_names() method in the ml_uvm library.**
  This method enables connecting TLM ports regardless of the language in which they are declared.

  Note The ml_uvm *e* library is compiled into Specman and can be used as shown above without any additional import, setting, or command-line option for Specman.

- **Use the external_uvm_path() attribute to associate units with their SV counterparts.**
  This promotes modularity of your code. Refer to "Connecting the e Proxy Sequencer to the SV Stub" on page 164 for more discussion and another example.

4.6 SystemVerilog Class-Based over *e*

This section describes multi-language environments in which the internal implementation of a Universal Verification Methodology (UVM) verification component (UVC) is implemented in *e*, and the higher level is implemented in SystemVerilog, with both layers following the UVM guidelines. This use model is useful when an integrator or test writer wants to integrate an *e* UVC into a SystemVerilog verification environment, and be able to write tests in SystemVerilog.

Cadence provides tools such as the UVM Interface Generator (UIG) to automate much of the process of creating a SystemVerilog API to an existing *e* UVC. This approach offers several benefits:

- Native SystemVerilog UVM user interface for the UVC integrator.
- The SystemVerilog API supports all the UVC configurations.
- Knowing *e* or the *e* implementation of the UVC is not needed for working with the SystemVerilog API. It is however needed for the creation of the SystemVerilog API.
- The SystemVerilog API is UVM-compatible and hence can be integrated with any other UVM-compliant UVCs.

Figure 4-5 on page 172 shows an example of a UVC with a SystemVerilog API. The lower layer is implemented in *e* (the pre-existing *e* UVC). The layer above is based on SystemVerilog interfaces and classes, which serve as an API to the *e* layer. The top layer is the actual application-specific SystemVerilog verification environment.

Figure 4-5   UVC — SystemVerilog API over *e*

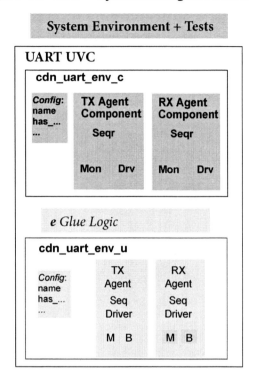

Created by verification engineers; UVC providers supply an example

Created by UIG/UIC, and included with the UVC

Created by the UVC developer, tightly coupled with the UIG generated code

Created by the e-UVC developer

To add SystemVerilog API to an *e* UVC, the *e* UVC developer must:

1. Generate the SystemVerilog API.

    Tools like the UVM Interface Generator (UIG) and UVM Interface Customizer (UIC) from Cadence automate much of the process of identifying which aspects of the *e* UVC should be exported and how, and then create the SystemVerilog wrapper and any required *e* glue logic.

2. Ship the *e* UVC with the SystemVerilog API files.

    The *e* UVC developer includes the additional source files (both *e* and SystemVerilog) with the *e* UVC, along with a usage example in SystemVerilog.

To integrate the UVC (with the SystemVerilog API) into a SystemVerilog verification environment, the SystemVerilog integrator must:

1. Instantiate and configure the SystemVerilog components exported by the *e* UVC.

2. Drive data items and sequences through the sequencer(s) exported by the *e* UVC.

3. Add module/system-level checking and coverage using the events and information in the monitors exported by the *e* UVC.

4.6.1 Simulation Flow in Mixed *e* and SystemVerilog Environments

The UVC created using this flow, being a mix of both *e* and SystemVerilog code, relies on a specific execution order between the SystemVerilog simulator and Specman. Because the top verification environment is in SystemVerilog, the SystemVerilog layer is the one that determines the configuration of the UVC. The SystemVerilog layer must therefore be allowed to run first, before the *e* layer is generated.

The order of events during simulation of an UVC that has an *e* core with a SystemVerilog API on top is the following:

1. Specman and the simulator are invoked, with the simulator's snapshot containing all of the compiled SystemVerilog code, and the Specman .esv file containing the compiled or loaded *e* UVC files.

2. The Verilog simulation begins, *before* Specman has gone through its setup and pre-run generation phases.

3. The SystemVerilog code creates instances of the SystemVerilog proxy classes which make up the SystemVerilog API, representing desired instances of the underlying exported *e* UVC units. Being components, all instances of the proxy classes must be created during the UVM *build* phase.

4. When the *build* phase in the UVM SystemVerilog environment completes, Specman resumes control and goes through its setup and pre-run generation.

5. The *e* glue logic within the UVC queries the SystemVerilog hierarchy and extracts all the required information on the instantiated proxy classes - their hierarchy and configuration.

6. The glue logic then creates an equivalent tree of *e* unit instances, all during pre-run generation.

7. Specman and the simulator then proceed with the other elaboration phases and the run phase, and simulation continues

4.6.2 Contacting Cadence for Further Information

Cadence provides the necessary tools, examples, and methodology documentation to facilitate the described use model. However, none of these are part of the UVM World contribution; they are part of the Cadence INCISIV release. To learn more about these capabilities, please contact Cadence.

4.7 UVM SystemC Methodology in Multi-Language Environments

This section introduces verification methodology for using the UVM SystemC library and the new features of the library. It describes using SystemC in multi-language environments, and contains the following sections:

Section	Contents
"Introduction to UVM SystemC" on page 174	Discusses SystemC use models and features in the UVM SystemC (UVM-SC) library.
"Using the Library Features for Modeling and Verification" on page 175	Discusses modeling configurable components with SystemC and using TLM ports for verification.
"Connecting between Languages using TLM Ports" on page 182	Shows how to connect a SystemC TLM DUT with a SystemVerilog testbench.
"Example of SC Reference Model used in SV Verification Environment" on page 188	Demonstrates one way of using the SystemC TLM as a reference model.
"Reusing SystemC Verification Components" on page 191	Suggests how to reuse SystemC verification components in a testbench.

4.7.1 Introduction to UVM SystemC

SystemC is primarily a modeling language. SystemC models can be used in various use models. The most important use models are:

- SystemC model used as a reference model for verification
- SystemC driver (BFM) or monitor reused in (SV/*e*) verification environment to access the DUT
- SystemC TLM model used to develop (SV/*e*) testbench, which can later be reused to verify the actual DUT
- SystemC TLM model used as stand-in component in RTL environment to replace a yet undeveloped piece of the DUT, or to accelerate simulation

The rest of this chapter focuses on the first and most important use model. "Reusing SystemC Verification Components" on page 191 suggests how to reuse SystemC verification components in a testbench.

4.7.1.1 New Features in UVM SystemC Library

The following features in the library are very valuable for creating transaction-level models for various use cases.

- Quasi-static components—components derived from uvm_component and thus inherit phasing and other built in features and can be configured and generated using the factory.

- Configuration—for quasi-static components generated using the factory, this feature provides configurability; certain attributes may be set before the component is generated by the factory and thus effect the generation.
- Multi-language connection of static and quasi-static SystemC ports and exports using TLM 1.0—allows simple and standard connection between the SystemC model and testbench written in other languages.
- Passing transaction data "by value" across SystemC language boundary—provides simple and standard communication for transactions derived from uvm_object.
- Matching SystemC/SystemVerilog TLM interfaces—provides seamless connection between model and testbench

Note that sc_modules can also be created at quasi-static time, and have the same phasing, but not factory generation capabilities. This maybe useful when having a transactor/adaptor component in SystemC that does not need all the UVM-SC bells and whistles.

For more details see the *UVM SystemC Library Reference* manual.

4.7.2 Using the Library Features for Modeling and Verification

There are several ways to take advantage of the new library features. This chapter deals with two major features:

- Modeling configurable components — The configuration mechanism allows us to create configurable models which provide flexibility for the verification engineer to align the behavior of the testbench and the model.
- Verifying using TLM ports — The TLM ports provide standard communication infrastructure for connecting the testbench to the model.

This chapter contains the following sections:

- "Modeling Configurable Component" on page 175
- "Verifying using TLM Ports" on page 177

4.7.2.1 Modeling Configurable Component

Creating a configurable transaction-level model has many advantages. When creating a SystemC model, it makes sense to make it as flexible as possible so it will be reusable in many projects.

There are different ways to make the model configurable, using derivation and compile time parameters. These methods require recompilation which is time consuming. Using the library, the model can be made configurable dynamically, without need for recompilation. In particular, the model and the verification environment can be aligned by applying the same configuration from the test.

Using SC_MODULE to Model the DUT

SC_MODULE is a static component which cannot be configured.

The simple example of SC_MODULE below shows a static component which cannot be configured:

```
SC_MODULE(producer) {
public:
  SC_CTOR(producer) { // constructor
    SC_THREAD(run);
    mode = 17;
  }
  void run() {
    wait(mode, SC_NS); // fixed behavior
    uvm_stop_request();
  }
private:
  int mode;
};
```

Making the Module Configurable

To create a configurable version of the same module, first derive it from uvm_component, and then create it using the uvm_factory.

Configurable component derived from uvm_component

To make the component configurable, derive it from uvm_component and use `get_config_*()` in the constructor. You should use `get_config_*()` for every field/parameter you want to make configurable. When the component is created, the configuration parameters obtained in the constructor can control the attributes and even the structure of the component.

```
class producer : public uvm_component {
public:
  producer(sc_module_name nm) : uvm_component(nm) { // constructor
    get_config_int("mode", mode);
  }
  UVM_COMPONENT_UTILS(producer)
  void run() {
    wait(mode, SC_NS); // Configurable behavior
    uvm_stop_request();
  }
private:
  int mode; // Configurable parameter
};
UVM_COMPONENT_REGISTER(producer)
```

Using uvm_factory to create the configurable component

To generate the configurable component, use the uvm_factory. The configuration is done via the `set_config_*()` function which can be called by a parent component (for example, the testbench) before the creation of this component. Wildcards are supported, so a family of producers can be configured at once.

```
class testbench : public uvm_component {
public:
  producer* prod;
  uvm_component* c;
  testbench(sc_module_name nm) : uvm_component(nm), prod() { }
  UVM_COMPONENT_UTILS(testbench)
  void build() {
    set_config_int("*", "mode", 7); // Configure the component
    c = uvm_factory::create_component("producer", "", "prod");
    prod = DCAST<producer*>(c);
    assert(prod);
  }
};
UVM_COMPONENT_REGISTER(testbench)
```

4.7.2.2 Verifying using TLM Ports

Using standard TLM ports and exports prepares the model to standard connections to the testbench, which may also be developed in a verification language such as *e* or SystemVerilog.

TLM ports are typed. An entire class (for instance, packet) can be transferred through the port.

A typical DUT has one or more input exports for drivers and one or more monitoring ports. The following sections suggest how to implement the verification interfaces.

As shown in Figure 4-6 on page 178 and Figure 4-7 on page 178, the DUT may be implemented as a separate SC_MODULE or as a configurable component under the testbench.

Figure 4-6 Configurable DUT Instantiated under the Testbench

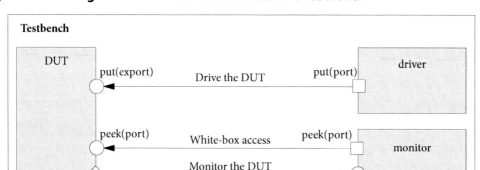

Figure 4-7 SC_MODULE DUT Separate from the Testbench

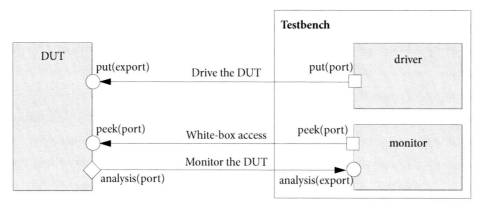

Example of using the DUT Ports

A typical DUT is driven by a driver, performs some activity and then outputs some result to a monitor. In addition, it may also provide access to internal status variables using a peek export.

The following example shows a simple DUT example focusing on the input and output ports. The DUT responds to the input after a short delay, sending the original packet to the output port. In addition, the "run" thread changes the value of the "status" independent of the packet processing activity.

```
class dut : public uvm_component,
      public tlm::tlm_blocking_put_if<packet>,
      public tlm::tlm_nonblocking_peek_if<int> {
public:
  sc_export<tlm::tlm_blocking_put_if<packet> > in; // input port
  sc_export<tlm::tlm_nonblocking_peek_if<int> > async; // async peek
  tlm::tlm_analysis_port<packet> out ; // output analysis port

  dut(sc_module_name nm) : uvm_component(nm), // constructor
```

```
                         in("in"), out("out"), async("async") {
    in(*this); // bind export to itself
    async(*this);
    status = 0;
  }
  UVM_COMPONENT_UTILS(dut)

  void put(const packet& t) { // Implement put export
    wait(5, SC_NS); // process the packet
    out.write(t); // send the packet out through analysis port
  }
  virtual bool nb_peek( int &t ) const { // Implement peek export
    t = status;
    return true;
  }
  virtual bool nb_can_peek( tlm::tlm_tag<int> *t ) const {}
  virtual const sc_event &ok_to_peek( tlm::tlm_tag<int> *t ) const {}

  void run() { // independent thread changing the status
    for (;;) {
      wait(7, SC_NS);
      status++;
    }
  }
private:
  int status;
};
UVM_COMPONENT_REGISTER(dut)
```

The input export "in" is blocking to allow the driver to synchronize with the DUT when the input is consumed. It is implemented by the put() method.

The output port "out" is an analysis port which requires that the monitor will react immediately, so it does not interfere with the DUT operation.

The peek export "async" allows the monitor to poll the value of the status at any time. The implementation is in the nb_peek() method.

Note All TLM ports are typed. The input and output ports expect a packet while the peek export gives access to an integer.

Driving the DUT

To drive the DUT, a driver component is needed. This example shows driving by a SystemC driver, but since the TLM ports support multi-language communication, the driver could be implemented in *e* or SystemVerilog as well.

The simple driver below generates packets and sends them to the DUT. A blocking put port at the input of the DUT will block until the data is consumed.

```
class driver : public uvm_component {
public:
  sc_port<tlm::tlm_blocking_put_if<packet> > out; // output to DUT
  driver(sc_module_name nm) : uvm_component(nm), out("out") {}
  UVM_COMPONENT_UTILS(driver)

  void run() {
    for (int i = 1; i <= 5; i++) {
      wait(10, SC_NS);
      packet* p = new packet(i*i);
      out->put(*p); // Drive the packet to the DUT
    }
    wait(100, SC_NS);
    uvm_stop_request();
  }
};
UVM_COMPONENT_REGISTER(driver)
```

Note The example shows uvm_component and takes advantage of the run() task which is automatically activated as the main thread.

Monitoring a Transaction-Level Model

Monitoring a transaction-level model may introduce several challenges:

- The monitoring interface of the model is not always well defined and if not available, the model must be modified to provide the results to the testbench monitor.
- White-box access into the model is not always provided; special accessor methods may be needed to allow the monitor visibility into the model.
- Timing may not be aligned between the reference model and the DUT, so the result may arrive before or after it is expected by the monitor.

Different monitoring tasks may be required to collect white- and black-box information from the transaction-level model:

- Black-box monitoring
  - The model puts the output on a port for the monitor; the port can be an analysis port or a put port.
- White-box synchronous
  - Monitoring is initiated from the monitor, for example, for comparing the internal model state with the expected state; the port can be a peek or get port implemented in the model.
- White-box asynchronous
  - Monitoring is initiated from the model, for example, to signal an internal event to the testbench; the port should be an analysis port.

This example below shows a simple monitor implemented by a uvm_component.

```
class monitor : public uvm_component,
                public tlm::tlm_analysis_if<packet> {
public:
  sc_export<tlm::tlm_analysis_if<packet> > in; // analysis port
  sc_port<tlm::tlm_nonblocking_peek_if<int> > async; // async input

  monitor(sc_module_name nm) : uvm_component(nm), in("in") {
    in(*this); // bind export to itself
  }
  UVM_COMPONENT_UTILS(monitor)

  void write(const packet& t) { // implement analysis port
    cout << sc_time_stamp() << ": received " << t << endl;
  }

  void run() { // independent thread polling the DUT status
    for (;;) {
      wait(10, SC_NS);
      int ret = async->nb_peek(data); // Poll white-box data
    }
  }
private:
  int data;
};
```

The example above shows an analysis port "in" where the monitor receives the output from the DUT. The implementation is in the write() method.

The peek port "async" used by the monitor to poll the status of the DUT independent of the DUT activity. The monitor obtains the status information by calling async->nb_peek(data);

Note All mandatory virtual functions must be implemented.

Connecting the DUT

The port binding is done on a higher level after the components are created during the build() phase.

```
class testbench : public uvm_component {
public:
  driver *drv_p;
  monitor *mon_p;
  dut *dut_p;
  uvm_component* c;
  testbench(sc_module_name nm) : uvm_component(nm), // constructor
         drv_p(0), mon_p(0), dut_p(0) { }
  UVM_COMPONENT_UTILS(testbench)
  void build() {
    c = uvm_factory::create_component("driver", "", "prod");
    drv_p = DCAST<driver*>(c);
    assert(drv_p);
```

```
        c = uvm_factory::create_component("monitor", "", "cons");
        mon_p = DCAST<monitor*>(c);
        assert(mon_p);
        c = uvm_factory::create_component("dut", "", "DUT");
        dut_p = DCAST<dut*>(c);
        assert(dut_p);
        drv_p->out(dut_p->in); // bind ports
        dut_p->out(mon_p->in);
        mon_p->async(dut_p->async);
    }
};
UVM_COMPONENT_REGISTER(testbench)
```

4.7.3   Connecting between Languages using TLM Ports

The small example below shows how to connect a SystemC TLM DUT with a SystemVerilog testbench. This example shows similar testbench-DUT connections to the previous example, except that the connections are multi-language connections.

Figure 4-8   Multi-Language Port Connection

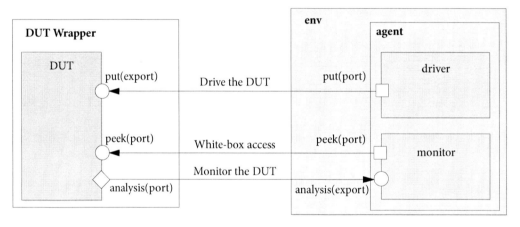

The environment is constructed from two tops: **dut_wrapper** in SystemC containing the DUT and **env** in SystemVerilog containing an agent with a driver and monitor.

4.7.3.1   Driving and Monitoring a TLM Model

The example in Figure 4-8 demonstrates the basic principles of connecting a SystemC TLM to a SystemVerilog testbench. The example shows:

- How to define the data type (packet)
- How to drive the DUT model from the testbench
- How to monitor the output from the DUT model
- Multi-language UVM considerations

Refer to the following sections:

- "Passing Data between Languages" on page 183
- "Driving the DUT" on page 184
- "Monitoring the DUT" on page 186

Passing Data between Languages

To be able to transfer a packet between SystemC and SystemVerilog compatible classes must be created in both languages, derived from uvm_transaction in SystemVerilog and uvm_object in SystemC. For example if there is a simple packet containing a kind (enum value), a name (string), a 10-bit integer and a fixed length array of integers the following class in SystemVerilog must be created:

```
class packet extends uvm_sequence_item;
  rand kind_t    kind;
  string         name;
  rand bit[9:0]  add;
  rand int       data[];
  `uvm_object_utils_begin(packet)
    `uvm_field_enum(kind_t, kind, UVM_DEFAULT)
    `uvm_field_string(name, UVM_DEFAULT)
    `uvm_field_int(add, UVM_DEFAULT)
    `uvm_field_array_int(data, UVM_DEC)
  `uvm_object_utils_end

  function new(string name = "packet");
    super.new(name);
    data = new[5];
    name = "ABCDE";
    add = 63;
    kind = MSG;
    for (int i=0; i<5; i++) data[i] = 0;
  endfunction : new
endclass
```

Note Using the uvm_object_utils and uvm_field_* automation macros on the fields, takes care of the necessary pack/unpack functions.

In SystemC, one must implement the required five virtual functions including the pack/unpack functions matching those from SystemVerilog:

```
class packet : public uvm_object {
  enum kind_t {MSG, DAT};
public:
  UVM_OBJECT_UTILS(packet)
  packet() {for(int i=0; i<5; i++) data[i] = 0;name = "abc";add = 63;}
  virtual ~packet() { }

  virtual void do_print(ostream& os) const {
    os << "kind: ";
```

```cpp
      if (kind == 0)
        os << "MSG";
      else
        os << "DAT";
      os << " name: " << name;
      os << " add: "  << add;
      os << " data: " ;
      for (int i=0; i<5; i++) os << data[i] << " " ;
      os << endl;
    }
    virtual void do_pack(uvm_packer& p) const {
      p << kind;
      p << name;
      p << add;
      p << 5; // Array length
      for (int i=0; i<5; i++) p << data[i];
    }
    virtual void do_unpack(uvm_packer& p) {
      int len;
      int kind_i;
      p >> kind_i;
      kind = (packet::kind_t)kind_i;
      p >> name;
      p >> add;
      p >> len; // Array length
      for (int i=0; i<5; i++) p >> data[i];
    }
    virtual void do_copy(const uvm_object* rhs) {...}
    virtual bool do_compare(const uvm_object* rhs) const {...}
  private:
    kind_t kind;
    std::string name;
    sc_bv<10> add;
    int data[5];
  };
  UVM_OBJECT_REGISTER(packet)
```

Notes

- The class must be registered by the UVM_OBJECT_REGISTER macro and the UVM_OBJECT_UTILS macro must be used.

Driving the DUT

This example assumes that the DUT model has a blocking put export at the input and an analysis port output, both of the same "packet" type.

If a legacy model has a different kind of input export, the model should be wrapped by a SystemC wrapper which implements the put export for the producer and transfers the transaction to the model.

Input to the Transaction-Level Model

The example shows a reference model with a blocking put export:

```
SC_MODULE(dut) ,public tlm::tlm_blocking_put_if<packet> {
public:
   sc_export<tlm::tlm_blocking_put_if<packet> > export_in;
   tlm::tlm_analysis_port<packet> port_out ;// output analysis port
   SC_CTOR(ref) : export_in("export_in"), port_out("port_out") {
     export_in(*this);
     ml_uvm::ml_uvm_register(&export_in);
     ml_uvm::ml_uvm_register(&port_out);
}
   void put(const packet& data_in) {…}
```

Note that the port is registered as a multi language port.

Driving from the SystemVerilog Testbench

The DUT is driven from the SystemVerilog testbench through the put port. The example shows a simple driver creating a packet and driving it into the DUT:

```
class driver extends uvm_driver;
   uvm_blocking_put_port #(packet) port_out;
   `uvm_component_utils(driver)

   function new(string name, uvm_component parent=null);
     super.new(name,parent);
     port_out = new("port_out", this);
   endfunction

   task run ();
     // Create packet p
     port_out.put(p);
   endtask : run
endclass
```

The SystemVerilog port is connected to the SystemC export, for example, in the SystemVerilog testbench:

```
function void connect_phase(uvm_phase phase);
    ml_uvm::connect(env.my_agent.bfm.port_out,
          "dut_wrapper.dut_i.export_in", "packet");
endfunction
```

Multi-Language UVM

Monitoring the DUT

The DUT below shows a transaction-level model used as a reference model. A SystemVerilog monitor is attached to the ports to monitor the reference model and get the expected results to be compared with the actual DUT results.

Monitoring the Transaction-Level Model

The reference model in this example has an analysis port for broadcasting the result packet. In addition, a peek port allowing asynchronous access to internal data in the reference model is added:

```
SC_MODULE(dut)  ,public tlm::tlm_blocking_put_if<packet>
                ,public tlm::tlm_nonblocking_peek_if<packet> {
public:
  sc_export<tlm::tlm_blocking_put_if<packet> > export_in;
  sc_export<tlm::tlm_nonblocking_peek_if<packet> > async; // async peek
  tlm::tlm_analysis_port<packet> port_out ;// output analysis port
  SC_CTOR(dut) : export_in("export_in"), async("async"),
                 port_out("port_out") {
    export_in(*this);
    async(*this);
    ml_uvm::ml_uvm_register(&export_in);
    ml_uvm::ml_uvm_register(&port_out);
    ml_uvm::ml_uvm_register(&async);
  }
private:
  sc_event dummy;
  packet data_out;
```

The code example shows that the DUT has an analysis port to broadcast the output packet. In addition it implements a peek port that provides the status information when the testbench needs it.

Note The peek port must transfer an object derived from uvm_object/uvm_transaction. It is not possible to peek a simple type like an integer.

Monitoring from the SystemVerilog Testbench

The reference model above is monitored by a SystemVerilog monitor in the testbench. Typically it will accept the output packet from the reference model and broadcast it through an analysis port to the rest of the verification environment:

```
class monitor extends uvm_monitor;
  uvm_analysis_imp#(packet, monitor) res_in;
  uvm_analysis_port#(packet) mon_bcast;
  `uvm_component_utils(monitor)

  function new(string name, uvm_component parent=null);
    super.new(name,parent);
    res_in = new("res_in",this);
    mon_bcast = new("mon_bcast", this);
  endfunction : new
```

Connecting between Languages using TLM Ports

```
  // Implement analysis export from reference model
  virtual function void write(packet p);
    // Perform checks and collect coverage
    mon_bcast.write(p); // broadcast to environment
  endfunction : write
endclass : monitor
```

The SV port is connected to the SC port, for example, in the SV testbench:

```
function void connect_phase(uvm_phase phase);
    ml_uvm::connect(env.my_agent.mon.mon_in,
        "dut_ref.dut_r.port_out", "packet");
endfunction : connect_phase
```

Multi-Language UVM Considerations

The SystemVerilog code must include uvm_macros.svh and import uvm_pkg and ml_uvm:

```
module topmodule;
  import uvm_pkg::*;
  import ml_uvm::*;
  `include "uvm_macros.svh"
```

The SystemC code must include ml_uvm.h and use the uvm namespace:

```
#include "ml_uvm.h"
using namespace uvm;
```

The environment is constructed from two tops: dut_ref in SystemC and top-module in SystemVerilog. The top level UVM component is the SystemVerilog testbench which is the verification environment. To simulate the environment use irun as follows:

```
irun -uvmtop SV:testbench \
-top dut \
-top topmodule \
-sysc \
-tlm2 \
${CDS_INST_DIR}/tools/uvm/sv/src/uvm_pkg.sv \
${CDS_INST_DIR}/tools/uvm/ml/ml_uvm_pkg.sv \
dut.cpp \
testbench.sv
```

Note A unique instance of the DUT is created as a top level component. If more than one instance is required, the type must be duplicated and renamed manually.

4.7.4 Example of SC Reference Model used in SV Verification Environment

One way of using the SystemC TLM as a reference model is demonstrated by the example in the following sections. A module/system UVC is used to drive the DUT and monitor its behavior. The monitor in this UVC gets access to the reference model and uses it to predict the DUT output for each input.

- The input monitor creates the abstract (TLM) input packet either from the wires or gets it from the driver.
- The input packet is made available to the module UVC monitor.
- The monitor sends the input packet to the reference model.
- The reference model computes and returns the expected result.
- The monitor sends the expected result to the add port of the scoreboard.
- The result is collected by the output monitor.
- The abstracted (TLM) output packet is made available to the module UVC monitor.
- The monitor sends the output packet to the match export of the scoreboard.

Figure 4-9 Verification with Reference Model

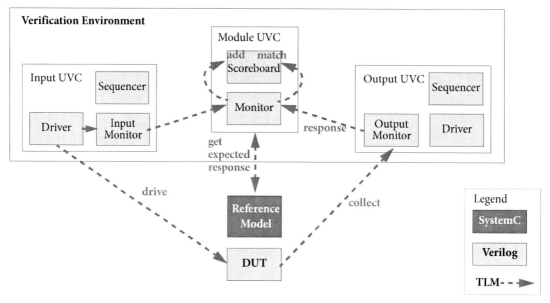

The connections to the reference model are similar to what was explained in "Connecting between Languages using TLM Ports" on page 182.

4.7.4.1 Selecting the Right Connection

The example in Figure 4-9 on page 188 can be implemented using different ports. There may be several options from which the integrator must choose.

Input to the Reference Model

A model that was developed for architectural analysis is likely to have a blocking port input. However, it is also possible that the model was developed as an untimed model and therefore its input export is non-blocking. The monitor must be able to handle both cases, so the input to the reference model must come from a task and not from a function.

Getting the Expected Result from the Reference Model

Different connections could be used to get the expected result from the reference model. For an untimed model, you can use uvm_non_blocking_transport, but it will not work for PVT models. Using a blocking put as the input and a non_blocking put as the output, both PV and PVT operation modes can be supported. This choice of ports is also likely to fit the original model, avoiding the need to implement a special wrapper for reusing the model.

White-Box Monitoring

The white-box monitoring port could possibly be non_blocking get, peek, or get_peek. The right choice depends on the implementation. In this case, assume that this export is not readily available in the model and must be added by the integrator.

4.7.4.2 Issues with using Reference Models

Reference models may use different levels of abstraction both in data and timing.

- Untimed (PV) models return the expected result immediately, so it always precedes the actual result.
- Loosely timed (PVT) models can provide the correct result but not exactly at the right time. The scoreboard must allow adding an expected result before or after the matching actual result.
- Cycle accurate models are expected to generate the output exactly when the DUT does. In this case, make sure that the expected result from the reference model gets added to the scoreboard before the corresponding output from the DUT.

Another concern may be ambiguity in the DUT specification. If, in a certain state-input combination, the DUT can choose one of several valid behaviors, the reference model may not be aligned with the DUT.

4.7.4.3 Using Bi-Directional Ports

The connection from the monitor to the reference model can be a bi-directional transport port. This port can send the input packet and receive the resulting packet in a single call.

```
tr.transport(pkt,rsp);
```

The transport call is blocking, and can therefore only be issued from a thread. The trigger for checking the reference model typically comes from an analysis port from a monitor. To work around this difficulty you must define an event that will trigger the reference model checking. The module level monitor may look like this:

Multi-Language UVM

```
class module_mon extends uvm_monitor;
   uvm_analysis_imp_add#(packet, module_mon) input_packet;
   uvm_analysis_imp_match#(packet, module_mon) actual_packet;
   uvm_analysis_port#(packet) mon_add;
   uvm_analysis_port#(packet) mon_match;
   uvm_blocking_transport_port #(packet,packet) tr;
   int data;
   uvm_event check_ref_event;

   packet pkt;
   `uvm_component_utils(module_mon)
   function new (string name, uvm_component parent=null);
     super.new(name,parent);
     mon_add = new("mon_add", this);
     mon_match = new("mon_match", this);
     tr = new("tr",this);
     check_ref_event = new("check_ref");
   endfunction : new

   function void build_phase(uvm_phase phase);
     super.build_phase(phase);
     input_packet = new("input_packet", this);
     actual_packet = new("actual_packet", this);
   endfunction : build_phase

   // Implement analysis ports
   virtual function void write_add(packet p);
     pkt = p;
     check_ref_event.trigger(); // indicate input packet received
   endfunction : write_add

   virtual function void write_match(packet p);
     mon_match.write(p); // match actual packet
   endfunction : write_match

   task run ();
     packet rsp;
     forever begin
       check_ref_event.wait_on(); // wait for arrival of input packet
       check_ref_event.reset();
       tr.transport(pkt,rsp); // Get expected result from ref model
       mon_add.write(rsp); // add expected packet
     end
   endtask : run
endclass : module_mon
```

4.7.5 Reusing SystemC Verification Components

Occasionally, SystemC verification components can be reused from the testbench. This requires additional port connections and substitution of the corresponding testbench components by proxy components that perform the translation and synchronization with the SystemC verification component.

Assume that you want to reuse a SystemC driver in a SystemVerilog testbench. The SystemC driver has a blocking put port to drive the DUT and gets the stimulus using a blocking get port from the verification environment.

Figure 4-10   Reusing a SystemC Driver

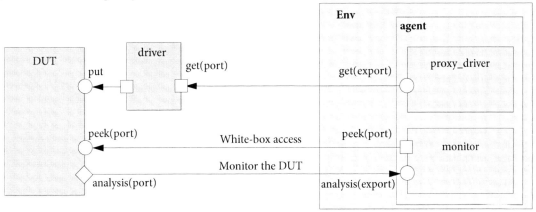

4.7.5.1   The SystemC Driver

The header file for the SystemC driver is as follows:

```
#include "ml_uvm.h"
#include "packet.h"
class driver : public uvm_component {
public:
  sc_port<tlm::tlm_blocking_put_if<packet> > out;
  sc_port<tlm::tlm_blocking_get_if<packet> > in;
  driver(sc_module_name nm) : uvm_component(nm),
                              in("in"), out("out") {
    ml_uvm::ml_uvm_register(&in);
    ml_uvm::ml_uvm_register(&out);
  }
  void run();
  packet get_next_item();   // Get next item from sequencer/proxy
  void drive_item(packet t);// Drive the item into the DUT
  UVM_COMPONENT_UTILS(driver)
};
UVM_COMPONENT_REGISTER(driver)
```

The actual implementation of the driver is in

- drive_item(packet t) - implementing the DUT driving
- run() - implementing the driver logic and obtaining items from the testbench

4.7.5.2 The Proxy Driver

To establish proper connection between the driver and the sequencer, one must instantiate a proxy driver in testbench. The proxy supports both the sequencer-driver protocol on the SystemVerilog side and the driver-proxy protocol with the SystemC driver.

The testbench modification is as follows:

- Derive a proxy driver from uvm_driver
- Define a blocking get export in the proxy and implement it
- Connect the ports

An example of proxy driver code is as follows:

```
class driver extends uvm_driver #(packet);
    uvm_blocking_get_imp #(packet, proxy_driver) proxy_in;
    `uvm_component_utils(proxy_driver)

    function new(string name, uvm_component parent=null);
      super.new(name,parent);
      proxy_in = new("proxy_in",this);
    endfunction : new

    function void build_phase(uvm_phase phase);
      super.build_phase(phase);
      ml_uvm::external_if(proxy_in, "packet");
    endfunction : build_phase

    virtual task get(output packet t);
        seq_item_port.get_next_item(t);
        seq_item_port.item_done(rsp);
    endtask : get
endclass : proxy_driver
```

This driver comes instead of the normal driver. It communicates with the SystemVerilog sequencer using the seq_item_port inherited from the base class. The activity of the proxy is implemented in the get export triggered by the SystemC driver, while the run task, which would normally be used for the driver implementation, is left empty.

4.7.5.3 Connecting the Ports

The ports are connected using the multi-language connect function as follows:

```
class testbench extends uvm_env;
  function void connect_phase(uvm_phase phase);
    ml_uvm::connect_names("driver.out", "dut.in");
    ml_uvm::connect_names("driver.in",
          "my_test.tb.uvc.agent.driver.proxy_in");
  endfunction : connect_phase
endclass : testbench
```

This should be done in the testbench level because these specific connections relate to a specific testbench with proxy driver replacing the normal driver.

4.8 Summary

The time when a single development team could design all the IP in and SoC is in the past. Each SoC is comprised of hundreds of individual IP blocks, each of which requires specialized expertise to design and verify. While the distribution of work is critical, if it isn't handled properly, integration can be a nightmare. The problem is that each IP development team will determine the level of reuse based on their functional commitments to the SoC. For example, a legacy block may be 100% reused because stability and consistency is paramount, but a high-performance block may be 100% new to achieve its specified goals. As a result, each team has to make its own decisions about internal coding and verification while adhering to some common APIs to enable integration. Any other decision, such as forcing unification at all levels, introduces quality and schedule risks that are unacceptable in today's rapid product cycle.

The UVM is the API that enables integration of IP and VIP build with IEEE standard languages. The UVM is architected to support multiple languages with its unique identification of sequence, driver, and monitor elements within the UVM agent and the use of transaction-level modeling (TLM) channels for communication. The transaction-level modeling was selected so that the VIP could operate at a unified, high-performance abstraction and connect to either HDL or SystemC models through the driver. Doing so enables design IP teams to select the language that best serves their needs—HDL for legacy and/or detailed models versus SystemC for high-performance and/or software-ready models. Stepping back to overall structure of the methodology, the UVM enables VIP teams to all select the language—typically IEEE standard *e*, SystemVerilog, or SystemC—that meets their scalable and reusable modeling requirements.

The globally developed SoC is plainly and simply a reality. If IP contributors are freed to choose the best language for their internal development, the SoC will have the best features and those will be of individually high quality. If they also choose to utilize the same API, then the SoC will be integrated quickly and achieve high overall quality. Because the Accellera UVM is architected for multiple languages while maintaining a consistent API, it has the potential to be a solution for SoC. The UVM multi-language implementation enables UVM to achieve that potential.

5 Developing Acceleratable Universal Verification Components (UVCs)

This chapter discusses the following topics:

- Introduction to UVM Acceleration
- UVC Architecture
- UVM Acceleration Package Interfaces
- SCE-MI Hardware Interface
- Building Acceleratable UVCs in SystemVerilog
- Building Acceleratable UVCs in e
- Collector and Monitor

5.1 Introduction to UVM Acceleration

The acceleratable Universal Verification Methodology (UVM) packages allow portions of a standard UVM environment to be accelerated using a hardware accelerator. The extended UVM acceleration packages include support for SystemVerilog and the *e* high-level verification languages (HVLs). Though this chapter only discusses UVM, acceleration for both the Open Verification Methodology (OVM) and UVM are supported. So, any references to UVM equally apply to OVM.

The purpose of extending UVM to include hardware acceleration is to enable the verification environment to execute faster. Hardware acceleration can dramatically increase run time performance and, therefore, allow more testing to be done in a shorter amount of time, and making the verification engineer more productive.

Although the main purpose of using the UVM acceleration library is to allow a hardware accelerator to be used, it is not restricted to hardware acceleration alone. UVM acceleration is truly an extension of the standard simulation-only UVM, and is fully backwards compatible with it. This means that Universal

Verification Components (UVCs) architected to be acceleratable can be used in either a simulation-only environment or a hardware-accelerated environment. However, UVCs that were not architected to leverage hardware acceleration will require some modifications to enable them to be used in a hardware-accelerated environment.

5.2 UVC Architecture

This section briefly describes the standard UVC architecture and explains how this differs from the acceleratable UVC architecture.

5.2.1 Standard UVC Architecture

UVCs based on the standard UVM typically contain the following three main components, which are themselves contained within an agent component, as shown in Figure 5-1 below. Each agent contains:

- A sequencer (also known as sequence driver)
- A driver (also known as BFM)
- A monitor

Figure 5-1   Standard UVC Architecture

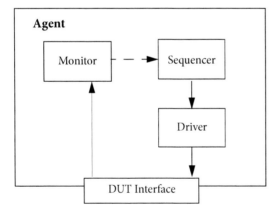

5.2.2 Active Agent

The architecture shown in Figure 5-1 is typical of an agent that actively drives stimulus into the device under test (DUT).

Stimulus is provided by the sequencer in an abstract form known as a *data item*. Data items are transactions that only contain stimulus information; the interface and protocol details related to the DUT are abstracted out. Data items in SystemVerilog are classes extended from the uvm_sequence_item class. In *e*, these are extended from the any_sequence_item item.

The driver connects to the DUT interface and applies the data items provided by the sequencer to this interface in accordance with the interface protocol.

A monitor is used to observe the activity on the DUT interface as well as activity on internal nodes of the DUT to collect coverage metrics about what parts of the DUT have been exercised. A standard UVM monitor usually includes a hard-coded connection to the interface as well as the coverage-collection functionality. Having a hard-coded connection to the interface is not ideal if the UVC is to be used to verify a DUT at multiple levels of abstraction because a new monitor will need to be created for each abstraction level.

5.2.3   Passive Agent

A UVC can be configured solely to collect DUT activity rather than to stimulate activity. The collected information can then be used by checkers, coverage tools, and the testbench itself for cases where up-to-date status is required. This is a typical scenario when the DUT is integrated into a system. Under these circumstances, the sequencer and driver components are disabled leaving only the monitor. The agent in this scenario is referred to as a *passive agent*.

Coverage information allows the verification team to ensure that the DUT is thoroughly tested by measuring the features that have been exercised, and the ones that have not. Coverage information can also be used by a scoreboard component that can be used to track the features that have been tested.

A UVC can be used to verify models at various levels of abstraction, each with different types of interfaces. Decoupling the stimulus generation from driving the physical DUT interface allows stimulus to be reused for verifying different abstractions of a given model by simply selecting the appropriate driver. This is most applicable to simulation environments that support the broadest range of HVL constructs.

5.2.4   Acceleratable UVCs

Acceleratable UVCs benefit from a slightly different architecture than simulation-only UVCs in order to maximize the performance gain provided by the hardware accelerator. Therefore, in order to describe acceleratable UVCs, a brief introduction to hardware acceleration must be given. More information about hardware acceleration can be found in the *UXE User's Guide*, which is included with the Cadence Palladium XP family of hardware accelerators.

5.2.4.1   Hardware Acceleration

Hardware acceleration is performed by combining a software simulator that executes on a workstation with a dedicated hardware-acceleration machine. The complete verification environment is partitioned to have some models executed by the simulator and others by the hardware accelerator. Models described using high-level verification language (HVL) constructs are executed by the simulator, and are said to reside in the *HVL partition*. Models described using hardware description language (HDL) constructs are executed by the hardware accelerator, and are said to reside in the *HDL partition*.

Hardware accelerators can only accelerate models that have been described using the acceleratable subset of an HDL. This subset is usually assumed to be the same as the register-transfer-level (RTL) subset defined for hardware synthesis, but this is often not the case. Hardware acceleration platforms usually accept a number of behavioral constructs as well as synthesizable constructs. So, the level of support is greater than that contained in the synthesizable subset of constructs; however, it is still a subset of the complete HDL. Any component that cannot be modeled using this subset must remain in the HVL partition. One requirement that must be fulfilled is that all models in the HDL partition must have signal-level interfaces. However, signal-level connections joining components in the HVL partition to components in the HDL partition are not efficient for achieving high runtime performance. Instead, a transaction-based connection must be used. The industry recognized this concept as being a key requirement in order to achieve high runtime performance when connecting a software simulator to a hardware accelerator. This led to the creation and standardization of the Accellera Standard Co-Emulation API: Modeling Interface, more commonly referred to as SCE-MI.

The use of a transaction-based interface between the software simulator and the hardware accelerator not only allows the communication between the two engines to be made more efficient, it also allows the execution of simulation models to be made more efficient. This is because simulation performance is reduced when the models being executed become more detailed and require timing. Therefore, simulating untimed models at the transaction level improves simulation performance.

The partitioning of transaction-level components and cycle-accurate signal-level components between the software simulator and hardware accelerator respectively, leads to a change in the overall verification environment architecture. Two separate top levels of hierarchy are created for each of the two partitions, with all communication between the two partitions being performed at the transaction level. Components like scoreboards, sequencers and monitors are placed in the HVL partition, while components like clock generators, reset generators, and the signal-level DUT are placed in the HDL partition, as shown in Figure 5-2.

Figure 5-2 Components of the HVL and HDL Partitions

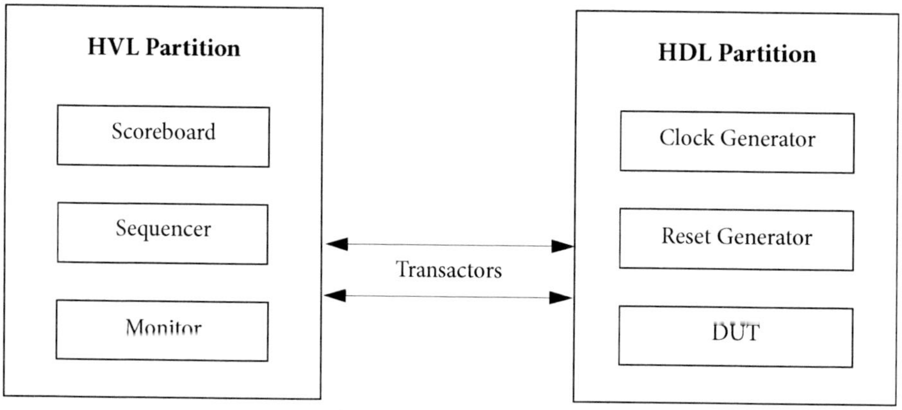

Acceleratable UVC Architecture

To enable high runtime performance to be achieved, all models that reside in the HVL partition should execute at the transaction level, and all models that require cycle-accurate timing should reside in the HDL partition. Transactors are used to allow components within each partition to communicate with each other efficiently. However, the architecture shown in Figure 5-1 on page 196 does not allow a clean division of functionality to be made because the monitor operates at the same abstraction level as the DUT, which for acceleration would be at the signal level. To address this, the monitor should be split into two components, a monitor and a collector, as shown in Figure 5-3.

Figure 5-3 Acceleratable UVC Architecture

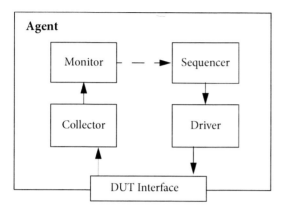

The purpose of the collector is to allow the physical interface required by the DUT to be separated from the functionality provided by the monitor. This means that the monitor and sequencer, and all hierarchical levels above, can operate at the transaction level, irrespective of the type of interface required by the DUT. Modeling these components at this level of abstraction is good for reuse as well as for increasing execution performance. The collector and driver components implement the physical interface required to enable the UVC to connect to the DUT, which can be easily altered depending on the type of interface required without affecting the rest of the UVC.

As mentioned previously, for the hardware acceleration mode, models that reside in the HVL partition operate at the transaction level, while those that reside in the HDL partition execute at the signal level. One consequence of configuring the UVC to use hardware acceleration is that the acceleratable collector and driver components must incorporate transactors to convert signal-level activity to transactions, and vice versa.

Acceleratable Transactors

Transactors are an abstraction bridge between the components that operate at the transaction level and the components that operate at the signal level. For hardware acceleration, transactors extend this capability by bridging between transaction-based components being executed by a software simulator and signal-based components being executed by a hardware accelerator as shown in Figure 5-4 on page 200.

Figure 5-4   Acceleratable Transactors

To bridge between the HVL partition and the HDL partition, the transactors have three main components:

- Proxy model

    The proxy model is instantiated in the HVL partition and accesses the communication channel by way of an Application Programming Interface (API).

- Bus Functional Model (BFM)

    The BFM is instantiated in the HDL partition and also accesses the communication channel by way of an API.

- Communication channel that connects between the Proxy and BFM

    Each channel is uni-directional and this is reflected in the choice of interface used within each of the partitions.

A simple transactor with one input and one output channel is shown in Figure 5-5.

Figure 5-5   Transactor Example with Input and Output Channels

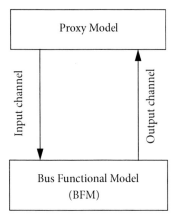

To enable transactors to operate on different vendor's hardware acceleration platforms, a standard vendor-independent API was defined and standardized by Accellera for connecting any software simulator to any hardware accelerator. The standard, known as the **Standard Co-Emulation API: Modelling Interface (SCE-MI),** defines a multichannel communication interface.

SCE-MI initially defined a macro-based interface. But later, it added a simpler Direct Programming Interface (DPI) and a more complex but feature-rich pipes-based interface. All of the above interfaces are described in the *Standard Co-Emulation API: Modelling Interface (SCE-MI) Reference Manual*, Version 2.0, or later, and is available from Accellera. The UVM Acceleration interface uses SCE-MI pipes communications channels.

SCE-MI pipes are unidirectional channels that allow transactions to be streamed from components in the HVL partition to components in the HDL partition and vice versa. A C language API is available to components residing in the HVL partition and a SystemVerilog API is available to components residing in the HDL partition.

To simplify the use of SCE-MI in the development of acceleratable UVCs, a UVM Acceleration library, *uvm_accel*, is provided in the Cadence UXE software release to hide the semantics of the SCE-MI C API presented to models that reside in the HVL partition. The uvm_accel package provides a UVM-based API that is native to the verification language being used. Most UVM verification environments are built using SystemVerilog, *e,* or a combination of both, and the uvm_accel library supports both. The uvm_accel package allows the proxy model part of a transactor to be written in SystemVerilog or *e*, whichever is the most suitable language, which is often the same as the language used to model the rest of the verification environment.

The BFM part of the transactor, implemented using the acceleratable subset of SystemVerilog and Verilog, uses the SCE-MI SystemVerilog interfaces to access the SCE-MI pipes based channels. These interfaces provide SystemVerilog tasks and functions that greatly simplifies the usage. More information about the SCE-MI Pipes interfaces can be obtained from the *Standard Co-Emulation API: Modeling Interface (SCE-MI) Reference Manual* from Accellera.

5.3 UVM Acceleration Package Interfaces

The UVM package provides two unidirectional interfaces, one to access input channels and the other to access output channels. The terms input and output are defined in relation to the hardware accelerator with input being into the hardware accelerator and output being out of the hardware accelerator.

For SystemVerilog, each interface is defined as a class that inherits from the uvm_accel_pipe_proxy_base. For *e*, each interface is defined as a unit that inherits from the uvm_accel_pipe_proxy_base unit.

5.3.1 uvm_accel_pipe_proxy_base Task and Function Definitions (SystemVerilog)

Task / Function	Definition
`extern function void build_phase(phase);`	Called during the environment `build_phase` phase. Gets configuration parameters such as `hdl_path` and `autoflush` and uses them to configure the proxy. To improve performance, you should run with `pipe_proxies` configured with autoflush disabled, whenever possible. For more information on autoflush, refer to the *Standard Co-Emulation API: Modeling Interface (SCE-MI) Reference Manual*.
`extern function void end_of_elaboration_phase(phase);`	Called at the end of elaboration. `hdl_path` must be configured before this function is called since port binding occurs during this phase.
`extern function void set_pipe_name(string name);`	Used to define the full hierarchical instance name of a pipe. The pipe name must be defined before `end_of_elaboration_phase` if it is to take effect. The name is given as a string.
`extern function string get_pipe_name();`	Returns the hierarchical instance name of the pipe that will be, or has been bound.
`extern function bit set_autoflush(bit enable);`	Sets the autoflush semantics of the pipe. An input of 1 turns autoflush on for all subsequent messages, and an input of 0 turns it off for subsequent messages. This setting can be made at anytime. The default is autoflush enabled (1).
`extern function bit get_autoflush();`	Returns the autoflush setting of the pipe.
`extern function int unsigned get_pipe_depth();`	Returns the number of elements that the pipe holds.
`extern function int unsigned get_pipe_width();`	Returns the number of bytes of each element.
`extern function int unsigned get_pipe_handle();`	Returns the internal handle that is used by the proxy to communicate with the actual pipe. This handle should not be used directly. Each actual pipe will have a unique handle.

SystemVerilog uvm_accel_input_pipe_proxy

The SystemVerilog `uvm_accel_input_pipe_proxy` class definition is shown below:

```
class uvm_accel_input_pipe_proxy #(type T=uvm_object,
```

```
      type S=uvm_accel_object_serializer#(T))       // Parameterizable
                                                    // serializer type
   extends uvm_accel_pipe_proxy_base;
   uvm_put_imp #(T,uvm_accel_input_pipe_proxy#(T)) put_export;  // TLM port
                                                    // binding
   uvm_analysis_port #(T) put_ap;                   // Analysis port
   extern function new(string name, uvm_component parent); // Constructor
   extern task put(T t);                            // Blocking put
   extern function bit try_put (T t);               // Non-blocking put
   extern function bit can_put();                   // Non-blocking can
                                                    // put test
endclass
```

Each input pipe proxy instance can be customized to accept different types of data item and use different serialization schemes. Customization is achieved via the parameters uvm_object and uvm_accel_object_serializer. Each data item is defined as a class in SystemVerilog which inherits from uvm_sequence_item. A data item typically contains data members that may or may not be randomized, UVM utility fields to enable or disable UVM automation for each of the data members, and constraints to constrain any data members that are to be randomized. In addition, serialization and de-serialization methods may also be provided for specific fields of the data item where the default serializer/de-serializer is not sufficient.

Table 5-1   SystemVerilog uvm_accel_input_pipe_proxy Task and Function Definitions

Task / Function	Definition
extern task put(T t);	Sends a user-defined data item of type T.
extern function bit try_put (T t);	Sends a user-defined data item of type T, if possible.
extern function bit can_put();	Returns 1 if the component is ready to accept the data item; 0 otherwise.

SystemVerilog uvm_accel_output_pipe_proxy

The SystemVerilog uvm_accel_output_pipe_proxy class definition is shown below:

```
class uvm_accel_output_pipe_proxy#(type T=uvm_object,
      type S=uvm_accel_object_serializer#(T))       // Parameterizable
                                                    // deserializer type
   extends uvm_accel_pipe_proxy_base;
   uvm_get_imp#(T,uvm_accel_output_pipe_proxy#(T)) get_export; // TLM port
      binding
   uvm_analysis_port#(T) get_ap;                    // Analysis port
   extern function new(string name, uvm_component parent);// Constructor
   extern task get(inout T t);                      // Blocking get
   extern function bit try_get (T t);               // Non-blocking get
   extern function bit can_get();                   // Non-blocking can get
endclass
```

Each output pipe proxy can be customized to accept different types of data item and use different de-serialization schemes. Customization is achieved by way of the parameters uvm_object and uvm_accel_object_serializer.

Table 5-2 SystemVerilog uvm_accel_output_pipe_proxy Task and Function Definitions

Task / Function	Definition
`extern task get(inout T t);`	Provides a new data item of type T.
`extern function bit try_get (T t);`	Provides a new data item of type T, if possible
`extern function bit can_get();`	Returns 1 if a new data item can be provided immediately upon request, 0 otherwise.

5.3.2   uvm_accel_pipe_proxy_base Task and Function Definitions (*e*)

e uvm_accel_input_pipe_proxy

Table 5-3   e uvm_accel_input_pipe_proxy Unit Definition

Task / Function	Definition
`get_pipe_full_path() : string`	Returns the hierarchical instance name of the pipe that will be, or has been bound.
`set_autoflush(enable : bool)`	Sets the autoflush semantics of the pipe. An input of 1 enables autoflush for all subsequent messages, and an input of 0 disables it for subsequent messages. This setting can be made at anytime. The default is autoflush enabled (1).
`get_pipe_autoflush() : bool`	Returns the autoflush setting of the pipe.
`get_pipe_depth() : uint`	Returns the number of elements that the pipe holds.

```
template unit uvm_accel_input_pipe_proxy of (<type>) like
uvm_accel_pipe_proxy_base {
    !value : <type>;
    m_in : interface_imp of tlm_put of <type> is instance; //TLM Interface
    put(value: <type> )                                     //Blocking put
    try_put(value: <type>) : bool                           //Non-blocking put
    can_put(): bool                                         //Non-blocking can
                                                            //put test
};
```

Each input pipe proxy instance can accept different types of data items. Each data item is defined as a unit in *e*, and is like any_sequence_item. A data item typically contains data members that may or may not be randomized, and includes constraints to constrain data members that are to be randomized. In addition, `pack` and `unpack` methods may also be provided for specific fields of the data item where the default packer or unpacker is not sufficient.

Table 5-4   e uvm_accel_input_pipe_proxy Task and Function Definitions

Task / Function	Definition
`put(value: <T>)`	Sends a user-defined data item of type T.
`try_put(value: <T>) : bool`	Sends a user-defined data item of type T, if possible.
`can_put(): bool`	Returns TRUE if the component is ready to accept the data item; FALSE otherwise.

e uvm_accel_output_pipe_proxy

The *e* uvm_accel_output_pipe_proxy unit definition is shown below:

```
template unit uvm_accel_output_pipe_proxy of (<type>) like
uvm_accel_pipe_proxy_base {
        !m_value : <type>;
        m_out : interface_imp of tlm_get of <type> is instance; // TLM Interface
        get(value: *<type>)                             // Blocking get
        try_get(value: *<type>): bool                   // Non-blocking
                                                        // get
        can_get(): bool                                 // Non-blocking
                                                        // can get
};
```

Each output pipe proxy can be customized to accept different types of data item.

Table 5-5   e uvm_accel_output_pipe_proxy Task and Function Definitions

Task / Function	Definition
`get(value: *<T>)`	Sends a user-defined data item of type T.
`try_get(value: *<T>): bool`	Sends a user-defined data item of type T, if possible
`can_get(): bool`	Returns 1 if a new data item can be provided immediately upon request, 0 otherwise.

5.4   SCE-MI Hardware Interface

The SCE-MI API used by the BFMs that exist in the HDL partition are defined in the *Standard Co-Emulation API: Modeling Interface (SCE-MI) Reference Manual*. The HDL side API for input and output interfaces are given here for reference. For complete details, refer to the SCE-MI Reference Manual.

5.4.1 SCE-MI Input Pipe Interface

```
interface scemi_input_pipe();
   parameter BYTES_PER_ELEMENT = 1;
   parameter PAYLOAD_MAX_ELEMENTS = 1;
   parameter BUFFER_MAX_ELEMENTS = <vendor specified>;
   localparam PAYLOAD_MAX_BITS = PAYLOAD_MAX_ELEMENTS * BYTES_PER_ELEMENT * 8;

   task receive(
       input int num_elements,                     // # elements to be read
       output int num_elements_valid,              // # elements that are valid
       output bit [PAYLOAD_MAX_BITS-1:0] data,     // data
       output bit eom );                           // end-of-message marker flag
       <implementation goes here>
   endtask

   function int try_receive(                       // return: #requested elements
                                                   // that are actually received
       input int byte_offset,                      // byte_offset into data
       input int num_elements,                     // # elements to be read
       output bit [PAYLOAD_MAX_BITS-1:0] data,     // data
       output bit eom );                           // end-of-message marker flag
       <implementation goes here>
   endfunction

   function int can_receive();                     // return: #elements that can
                                                   // be received
       <implementation goes here>
   endfunction

   modport receive_if(import receive, try_receive, can_receive );
endinterface
```

5.4.2 SCE-MI Output Pipe Interface

```
interface scemi_output_pipe();
   parameter BYTES_PER_ELEMENT = 1;
   parameter PAYLOAD_MAX_ELEMENTS = 1;
   parameter BUFFER_MAX_ELEMENTS = <vendor specified>;
   localparam PAYLOAD_MAX_BITS = PAYLOAD_MAX_ELEMENTS * BYTES_PER_ELEMENT * 8;

   task send(
       input int num_elements,                     // input: #elements    to be
   written
       input bit [PAYLOAD_MAX_BITS-1:0] data, // input: data
       input bit eom );  // input: end-of-message marker flag
       <implementation goes here>
   endtask

   task flush;
```

```
        <implementation goes here>
    endtask

    function int try_send(                      // return: #requested elements
                                                // that are actually sent
        input int byte_offset,                  // input: byte_offset into
                                                // data, below
        input int num_elements,                 // input: #elements to be sent
        input bit [PAYLOAD_MAX_BITS-1:0] data,  // input: data
        input bit eom );                        // input: end-of-message marker
                                                // flag
        <implementation goes here>
    endfunction

    function int can_send();                    // return: #elements that can be sent
        <implementation goes here>
    endfunction

    modport send_if( import send, flush, try_send, can_send );
endinterface
```

5.5 Building Acceleratable UVCs in SystemVerilog

5.5.1 Data Items

Data items are transactions, which are implemented as class objects that are inherited from uvm_sequence_item, that itself inherits from uvm_transaction. A data item contains data members, UVM utility fields to enable or disable UVM automation for each of the data members, and constraints to constrain any data members that are to be randomized. In addition, you may provide your own serialization and de-serialization methods. The code snippet below, taken from a simple SystemVerilog example, yamp, shows the class definition of a data item called yamp_transfer along with its data members.

```
    typedef enum bit { READ, WRITE } direction_t;   // Enumerated type used to
                                                    // define memory access
                                                    // direction
    class yamp_transfer extends uvm_sequence_item;  // yamp_transfer class
                                                    // inherited from
                                                    // 'uvm_sequence_item'
                                                    // class
        rand direction_t direction;                 // Memory access direction
                                                    //(READ OR WRITE)
        rand bit [2:0] wait_states;                 // Used by the Driver to
                                                    // insert wait states
        rand bit [3:0] transfer_delay;              // Used by the Driver to
                                                    // insert a transfer delay
        rand bit [7:0] size;                        // Size of data transfer
        rand bit [15:0] addr;                       // Start address of memory
                                                    // access
```

```
        rand bit [15:0] data [];                    // Data to be read or
                                                    // written to memory
```

SystemVerilog data members can be randomized as shown by preceding their declaration with the keyword `rand`. Data items can contain statically sized data members as well as dynamically sized data members such as `data[]` shown in the example.

Data items that contain randomly assigned data members require constraints to constrain the range of values they will be assigned. Constraints can be defined within the class definition as shown below or in a separate constraints file.

```
    constraint default_wr_size_c {(direction == WRITE) -> data.size() == size;
                                  (direction == READ) -> data.size() == 0; }
    constraint default_size_c { size inside { [1:10] }; }
    constraint default_delay_c { transfer_delay inside {[1:5]};}
```

`uvm_object_utils` macros are used to enable common operations declared in `uvm_object` such as copy, compare, and print as shown below.

```
    `uvm_object_utils_begin(yamp_transfer)            // Start of UVM utility
                                                      // macro definitions
    `uvm_field_enum(direction_t, direction, UVM_ALL_ON)
    `uvm_field_int(wait_states, UVM_ALL_ON)
    `uvm_field_int(transfer_delay, UVM_ALL_ON)
    `uvm_field_int(size, UVM_ALL_ON)
    `uvm_field_int(addr, UVM_ALL_ON + uvm_HEX)
    `uvm_field_array_int(data, UVM_ALL_ON + UVM_HEX + UVM_NOPACK)
    `uvm_object_utils_end                             // End of UVM utility
                                                      // macro definitions
```

In order to transfer a data item from the proxy in the HVL partition to the BFM in the HDL partition, the data members must be packed, or serialized, into a vector of bits as shown below.

Figure 5-6 Packed Implementation of Data Item yamp_transfer

dir	wait_states	transfer_delay	size	addr	data []

UVM provides packing capabilities which may or may not be suitable for the data item to be transferred. When data members are statically sized the standard packer is usually sufficient but alternative packing schemes may be required for dynamically sized data members if they have specific requirements. If a field is to be packed using a customized serializer the attribute UVM_NOPACK should be set using the `` `uvm_object_util_* `` macro. If the dynamic members do not have any specific requirements then the standard UVM packer can be used for static and dynamic data members. An example of specific `pack` function required by the `yamp` example is shown below.

```
    function void do_pack (uvm_packer packer);
        foreach(data[i]) packer.pack_field_int(data[i],16);
    endfunction
```

Data items received by the proxy in the HVL partition, from the BFM in the HDL partition, must be unpacked back into the data item class structure. It is the `unpack` operation that usually dictates whether custom pack and unpack functions are required. The reverse operation employed by the packer must be used by the unpacker. Therefore, if a customized packer was defined then a customized unpacker or deserializer must also be defined. The code snippet below shows the custom unpacker used by the `yamp` example.

```
function void do_unpack (uvm_packer packer);
   data = new [size];                          //size was automatically unpacked
   foreach(data[i]) data[i] = packer.unpack_field_int(16);
endfunction
```

5.5.2 Acceleratable Driver (SystemVerilog)

The *driver* is responsible for taking data items from the sequencer and driving them onto the DUT interface. The DUT can be modeled at multiple levels of abstraction. So, the driver must be able to accommodate each of the interfaces presented by each type of model. This not only affects the type of physical interface used it also affects the functionality of the driver itself. To be able to reconfigure the driver to operate at different levels of abstraction, an enumerated type `uvm_abstraction_level_enum` is used. This enumerated type is defined in the `uvm_accel` package provided by Cadence.

In SystemVerilog, the enumerated type is defined as follows:

```
typedef enum bit [1:0] {UVM_SIGNAL, UVM_TLM, UVM_ACCEL}
uvm_abstraction_level_enum
```

The values defined by this type configure the UVC to operate in pure simulation at the signal level (`UVM_SIGNAL`) or transaction level (`UVM_TLM`) or use hardware acceleration (`UVM_ACCEL`).

When configured for hardware acceleration an acceleratable transactor is used to bridge the gap between the components that operate at the transaction level, which are executed by the software simulator, and the components that operate at the signal level, which are executed by the hardware accelerator. This same acceleratable transactor can also be used for signal based simulation. However, UVCs that have been created for simulation typically use a virtual interface to connect the driver to the DUT and implement the BFM using behavioral constructs. This implementation can continue to be used for simulation to allow a gradual migration to hardware acceleration if required. When using the behavioral BFM the `uvm_abstraction_level_enum` should be set to `UVM_SIGNAL`. If the UVC is to be used to verify abstract SystemC TLM models, the `uvm_abstraction_level_enum` should be set to `UVM_TLM`. The behavior of the driver along with the interface it uses to connect to this model should be customized to suit this type of model.

The following code shows the SystemVerilog code that defines the part of the driver that resides in the HVL partition for the `yamp` example.

```
class yamp_master_driver extends uvm_driver #(yamp_transfer);
   // Virtual interface used to drive HDL signals
   virtual interface yamp_if vif;
   // UVM abstraction level
   protected uvm_abstraction_level_enum abstraction_level = UVM_SIGNAL;
   // SCE-MI input pipe interface
   protected uvm_accel_input_pipe_proxy#(yamp_transfer) m_ip;
   // SCE-MI output pipe interface
```

```
      protected uvm_accel_output_pipe_proxy#(yamp_transfer) m_op;
      // UVM build function
      extern virtual function void build_phase(uvm_phase phase);
      // UVM run task
      extern virtual task run_phase(uvm_phase phase);
      // Task used to drive signals in UVM_SIGNAL mode
      extern virtual protected task get_and_drive();
      // Task used to drive signals in UVM_ACCEL mode
      extern virtual protected task get_and_drive_accel();
endclass : yamp_master_driver
```

The yamp_master_driver inherits from the uvm_driver class and operates on a data item of type yamp_transfer. This example shows a virtual interface, vif, which is used for signal level simulation, and two uvm_accel pipe proxy interfaces, m_ip and m_op that are used for hardware acceleration.

Two uvm_accel pipe proxy interfaces are required for the yamp example since bidirectional communication is required. Each uvm_accel pipe proxy interface is unidirectional; therefore, the need for one input interface and one output interface. For most protocols, bidirectional communication is required so it is typical for two or more interfaces to be instantiated. Each uvm pipe proxy interface takes a data item type as a parameter.

Standard UVM tasks and functions must be defined for each driver. It is recommended that different tasks for each level of abstraction are defined rather than implementing the driver functionality in one task for all the supported levels of abstraction. In the yamp example, the get_and_drive() task implements the signal-level simulation driver functionality and the get_and_drive_accel() task implements the hardware-acceleratable driver functionality. Separating the code into distinct task makes the code easier to understand and debug.

5.5.2.1 build_phase(uvm_phase phase) Function

Each UVM component that inherits from the uvm_component class should provide an implementation for a build_phase function. Each build_phase function is called during the UVM build_phase simulation phase to construct the environment hierarchy. In the example shown below, the abstraction_level is used to determine the type of interface required by the driver.

```
function void yamp_master_driver::build_phase(uvm_phase phase);
   super.build_phase(phase);
   if (abstraction_level == UVM_ACCEL)
      begin
         m_ip = new("m_ip", this);    // Construct an input port
         m_op = new("m_op", this);    // Construct an output port
         uvm_config_db#(string)::set(this,"m_ip", "hdl_path", "inbox0");
                  // hdl_path used for input port binding
         uvm_config_db#(string)::set(this,"m_op", "hdl_path", "outbox0");
                  // hdl_path used for output port binding
      end
endfunction : build_phase
```

For hardware acceleration, the abstraction_level must be set to UVM_ACCEL to inform the driver to build and configure a transaction based interface. For the yamp example, two ports are constructed: an input port called m_ip and an output port called m_op. These ports must be bound to valid channels before they can be used and this is achieved by defining a string called hdl_path for each port.

Port binding is configured by calling the UVM uvm_config_db#(string)::set function for each port defined in the HVL partition. The uvm_config_db#(string)::set function causes configuration settings to be created and placed in the uvm_config database. The uvm_config_db#(string)::set function requires the name of the port instance in the HVL partition, the name of the string variable to be configured (which is hdl_path for port binding), and the full hierarchical path from the top level of the HDL partition down to the appropriate port instance in the HDL design hierarchy. In the example given, the full hierarchical path is defined by concatenating the m_hdl_path variable with the specific port instance name. The m_hdl_path variable is set by the test environment and is the hierarchical path from the top level of the HDL partition down to the BFM instance. The agent then appends the specific port instance name to this path.

If the HDL port, defined by hdl_path, is compatible with the HVL port, it will be bound during the end_of_elaboration_phase phase; if not, an error will occur. Therefore, the hdl_path for each port must be defined before the end_of_elaboration_phase phase; it is common to do this during the build_phase phase as shown.

5.5.2.2 run_phase(uvm_phase phase) Task

Each UVM component that inherits from the uvm_component class, should provide an implementation for a run_phase task. Each run_phase task is called during the UVM run_phase simulation phase and defines the behavior of the driver. The required functionality of the driver will differ depending on the level of abstraction used to implement the DUT. Therefore, the abstraction_level is tested and used to alter the driver's behavior as shown below.

```
task yamp_master_driver::run_phase(uvm_phase phase);
    if (abstraction_level == UVM_SIGNAL)       // Signal level simulation
        fork
            get_and_drive();                   // Drive signal level DUT interface
        join
    else if (abstraction_level == UVM_ACCEL)   // Hardware acceleration fork
        fork
            get_and_drive_accel();             // Drive SCE-MI transaction level
                                               // interface
        join
endtask
```

If the abstraction_level is set to UVM_SIGNAL, and a signal-level behavioral BFM has been created for simulation, which is typical of legacy UVCs, a get_and_drive() task should be called. This task implements the functionality required to drive this type of interface.

If the abstraction_level is set to UVM_ACCEL, a get_and_drive_accel() task should be called. Different tasks are defined for simulation and acceleratable drivers to allow a legacy behavioral implementation to be used, and coexist with an acceleratable implementation. Acceleratable drivers can be used with hardware acceleration or simulation. Therefore, the same task could be called, irrespective of whether the `abstraction_level` is set to `UVM_SIGNAL` or `UVM_ACCEL`. This is configured in the `run_phase` task.

5.5.2.3 get_and_drive() Task

The `get_and_drive()` task requests data items from the sequencer, and when appropriate drives the virtual DUT interface signals. It implements the signal-level protocol required by the DUT and drives the DUT signals directly, as shown in the code snippet from the `yamp` example below.

```
task yamp_master_driver::get_and_drive();
   if(vif.sig_reset!==0) @(negedge vif.sig_reset);
   forever begin
      @(posedge vif.clk);
      seq_item_port.get_next_item(req);           // Get new item from
                                                  // the sequencer
      if (transfer.direction == WRITE) begin      // Drive the virtual
                                                  // virtual interface
                                                  // signals
        vif.rd <= 0;
        for (int i=0;i < transfer.size; i++) begin
          repeat (transfer.wait_states) @(posedge vif.clk);
          vif.we <= 1;
          vif.di <= transfer.data[i];
          @(posedge vif.clk);
          vif.addr <= vif.addr + 1;
          vif.we <= 0;
      <rest of implementation>
      seq_item_port.item_done();                  // Communicate item done
                                                  // to the sequencer
   endtask : get_and_drive
```

5.5.2.4 get_and_drive_accel() Task

The `get_and_drive_accel()` task uses the `uvm_accel` interfaces to send and receive data items as transactions from the HVL partition into the HDL partition where a hardware BFM drives the DUT signals. The `get_and_drive_accel()` task does not implement any signal-level protocol functionality it operates purely at the transaction level. The HDL BFM is implemented as a separate module and is instantiated in the HDL hierarchy partition which will be described in the next section.

The `uvm_accel` ports use standard transaction-level modeling (TLM) semantics to send and receive transactions by way of SCE-MI communication channels. The code snippet below shows the blocking put and blocking get tasks being used to send and receive data items.

```
task yamp_master_driver::get_and_drive_accel();
   forever begin
      seq_item_port.get_next_item(req);    // Get new item from the sequencer
      m_ip.put(req);                        // Drive the item
      if(req.direction == READ) begin
```

```
            m_op.get(req);
            seq_item_port.item_done(req);
        end
        else begin
            //Communicate item done to the sequencer
            seq_item_port.item_done();
        end
        <rest of implementation>
    end
endtask
```

Once a data item has been taken from the sequencer, it can be put into an input channel using the blocking `put()` function associated with the port that is bound to that channel. The blocking `put()` function blocks until the transaction has been taken from the channel at the opposite end. This means that the `get_and_drive_accel()` task does not need to implement any sort of *wait* before informing the sequencer that the current sequence item has been done. This is simpler than in the non-accelerated case where you must implement any code required to allow one sequence to be completed before the next one is started.

5.5.2.5 Acceleratable Driver BFM (SystemVerilog)

The acceleratable driver BFM resides in the HDL partition and implements the signal level protocol functionality required to drive the DUT. The acceleratable driver BFM contains SCE-MI pipes interfaces which are bound to ports within the driver component that resides in the HVL partition. The `get_and_drive_accel()` task passes transactions through a SCE-MI pipe to the driver BFM which must extract the transaction and apply it to the DUT signal level interface.

The acceleratable driver BFM must be written in acceleratable SystemVerilog or Verilog for it to be accelerated by a hardware accelerator. The driver code should be partitioned into separate files to reflect code that is to be simulated and code that is to be accelerated. This simplifies the overall compilation process and makes the code easier to maintain.

Note The Cadence UVM Acceleration package provides *e* and SystemVerilog interfaces to allow access to the pipes on the HVL side. Therefore, the same acceleratable driver BFM can be used in both environments.

Each driver BFM must instantiate appropriate SCE-MI pipes ports to mirror those defined in the driver's proxy which exists in the HVL partition. If the ports at each end of the communication channel are not compatible, they will not be bound and elaboration will fail. The code snippet below, taken from the `yamp` example, shows a SCE-MI input pipe called `inbox0()` and a SCE-MI output pipe called `outbox0()`.

```
module yamp_master_driver_bfm (
    input     wire         clk,
    output    reg          cmd,
    output    reg[7:0]     len,
    output    reg          we,
    output    reg          ce,
    output    reg          rd,
    output    reg[15:0]    addr, di,
    input     wire[15:0]   dout,
```

```
      input    wire         scemi_mode
);                                           // SCE-MI input pipe instantiation
    scemi_input_pipe #(2, 1) inbox0 ();   // SCE-MI output pipe instantiation
    scemi_output_pipe #(2, 1) outbox0 ();
<rest of implementation>
```

Both `inbox0` and `outbox0` have the parameters BYTES_PER_ELEMENT set to 2 and PAYLOAD_MAX_ELEMENTS set to 1.

BYTES_PER_ELEMENT = 2 means that each message element received will contain two bytes.

PAYLOAD_MAX_ELEMENTS = 1 means that only one message element will be received at a time.

These two parameters define the width of the data that can be received by an input port or sent by an output port. Each ports width is defined by the parameter PAYLOAD_MAX_BITS that is defined as shown in the following formula:

```
PAYLOAD_MAX_BITS = PAYLOAD_MAX_ELEMENTS * BYTES_PER_ELEMENT * 8;
```

Therefore, the ports in the example above are capable of receiving or sending messages only 16-bits wide during each transfer.

A UVC can contain different types of driver to suit the level of abstraction used to model the DUT. If a simulation-based driver and an acceleratable driver have both been implemented, it is important to ensure that only one driver drives the DUT at any one time. The `abstraction_level_enum` should be used to define the value of `scemi_mode`. When the SCE-MI hardware acceleratable driver is to be used `scemi_mode` should be set to 1; for all other scenarios, `scemi_mode` should be set to 0. This is usually defined at the top level of the UVC. The code snippet below, taken from the `yamp` example, shows that the output `we_r`, `ce_r`, and `rd_r` are tri-stated, unless `scemi_mode` has been set to 1.

```
// Output tri-state logic
    always@(we_r or scemi_mode) we <= scemi_mode3we_r:1'bz;
    always@(ce_r or scemi_mode) ce <= scemi_mode3ce_r:1'bz;
    always@(rd_r or scemi_mode) rd <= scemi_mode3rd_r:1'bz;
    <rest of implementation>
```

The SCE-MI pipes HDL API provides blocking and non-blocking tasks and functions. The code snippet below, taken from the `yamp` example shows how the blocking `receive()` task is used.

```
always@ (posedge clk) begin
    if(scemi_mode) begin
        inbox0.receive(1, num_recv, idata, eom);
        {len_r, delc, ws, cmd_r} = idata;
<rest of implementation>
```

At the positive edge of the clock called `clk`, the `receive()` task associated with `inbox0` is called with the following arguments:

```
Num_elements = 1
Num_elements_valid = num_recv
Output_data = idata
EOM = eom
```

`Num_elements` defines how many elements are to be put into the variable idata when a transaction has been received. This example deals with one message element at a time. A transaction can contain many message elements, and the BFM designer needs to decide the most efficient implementation.

`Num_elements_valid` defines the number of received elements that are valid. This can be used by the BFM to determine the elements to be used when multiple elements are received in one transfer. This is not relevant in this example because only one element can be received at one time.

`Output_data` defines the variable in which received data will be written into. The width of this variable should be defined by `PAYLOAD_MAX_BITS` as described above.

`EOM` defines whether the message element received is a single message element or a part of a continuous stream of message elements. Using `EOM`, it is possible to send transactions that contain a variable number of message elements during each transfer. When `EOM` is set to `1`, the element received is the last element. When `EOM` is set to 0, there are more elements available to read.

More information about the SCE-MI hardware API can be found in the *Standard Co-Emulation API: Modeling Interface (SCE-MI) Reference Manual*.

5.6 Building Acceleratable UVCs in *e*

5.6.1 Data Items

Data items are transactions that are implemented as struct objects that derive from `any_sequence_item`. A data item contains data members, constraints to constrain any data members that are to be randomized, and methods for manipulating the data members or the struct itself. The code snippet below, taken from the *e* yamp example, shows the `struct` definition of a data item called `transfer_s` along with its data members.

```
struct transfer_s like any_sequence_item {      // transfer_s struct
    %direction     : yamp_direction_t;          // Memory access
                                                // direction
                                                // (READ OR WRITE)
    %wait_states   : uint (bits : 3);           // Used by the driver
                                                // to insert wait states
    %delay_clocks  : uint (bits : 4);           // Used by the driver
                                                // to insert a transfer
                                                // delay
    %size          : uint (bits : 8);           // Size of data transfer
    %addr          : yamp_addr_t;               // Start address of
                                                // memory access
    %data          : list of uint (bits : YAMP_DATA_WIDTH);   // Data to be
                                                // read or written to
                                                // memory
```

Data items that contain randomly assigned data members require constraints to constrain the range of values they will be assigned. Constraints can be defined within the struct definition as shown below or in a separate constraints file.

```
keep soft data.size() == size;
keep direction == WRITE  =>  data.size() == size;
keep soft size > 0;
keep soft size < 10;
```

To transfer a data item from the proxy in the HVL partition to the BFM in the HDL partition, the data members must be packed into a vector of bits as shown in Figure 5-7 below.

Figure 5-7 Packed Implementation of Data Item yamp_transfer

direction	wait_states	delay_clocks	size	addr	Data

e provides built-in `pack` and `unpack` methods to create a list of bits that is a concatenation of the members contained in the data item `struct`. The acceleratable driver must understand the packing scheme used in order to extract each member from the data item received.

5.6.2 Acceleratable Driver (*e*)

The UVC BFM is responsible for taking data items from the sequencer and driving them onto the DUT interface. The DUT can be modeled at multiple levels of abstraction. So, the BFM must be able to accommodate each of the interfaces presented by each type of model. This not only affects the type of physical interface used, it also affects the functionality of the BFM itself. To be able to reconfigure the BFM to operate at different levels of abstraction, an enumerated type `uvm_abstraction_level_t` is used. This enumerated type is defined in the `uvm_accel` package provided by Cadence.

In *e*, the enumerated type is defined as follows:

```
type uvm_abstraction_level_t : [UVM_SIGNAL, UVM_TLM, UVM_ACCEL]
                              (bits : 2);
```

The values defined by this type configure the UVC to operate at one of the following levels:

- Pure simulation at the signal level (`UVM_SIGNAL`)
- Pure simulation at the transaction level (`UVM_TLM`)
- Use hardware acceleration (`UVM_ACCEL`)

When configured for hardware acceleration, an acceleratable transactor is used to bridge the gap between the components that operate at the transaction level and the signal level. Transaction-level components are executed by the software simulator, and signal-level components are executed by the hardware accelerator. The same acceleratable transactor can be used in a simulation-only environment as well as with hardware acceleration. However, multi-purpose UVCs that are configured to operate in `UVM_SIGNAL` mode typically implement the BFM in behavioral *e* code. This implementation can continue to be used for simulation to allow a gradual migration to hardware acceleration, if required. When using the behavioral BFM, the `uvm_abstraction_level_t` should be set to `UVM_SIGNAL`. If the UVC is to be used to verify abstract SystemC TLM models, the `uvm_abstraction_level_t` should be set to `UVM_TLM`. The behavior of the driver along with the interface that it uses to connect to this model, should be customized to suit this type of model.

Acceleratable Driver (e)

One of the main features of *e* is that it provides aspect orientation. This means that objects can be extended to accommodate new functionality or manipulate existing functionality. For UVM Acceleration it is common for the different abstraction levels to be implemented by extending existing units.

The following code shows the *e* code which defines the part of the driver that resides in the HVL partition for the yamp example.

```
extend UVM_ACCEL master_bfm {
      keep hdl_path() == "xi0";                   // HDL path
      m_ip : uvm_accel_input_pipe_proxy of transfer_s is instance; // Input
                                                                   // Port
      keep m_ip.hdl_path() == "inbox0";
      m_op : uvm_accel_output_pipe_proxy of transfer_s is instance; // Output
                                                                    // Port
      keep m_op.hdl_path() == "outbox0";
      m_in :   interface_port of tlm_put of transfer_s is instance; // Input
                                                                    // Port
      m_out : interface_port of tlm_get of transfer_s is instance;  // Output
                                                                    // Port
      connect_ports() is also{                    // Port
                                                  // Binding
      m_in.connect(m_ip.m_in);
      m_out.connect(m_op.m_out);
      };
      drive_transfer (cur_transfer : transfer_s)  // Drive
                                                  // transfer
                                                  // method
};
```

The master_bfm extends the generic BFM and is extended further when the abstraction level is set to UVM_ACCEL. This example shows two uvm_accel pipe proxy interfaces, m_ip and m_op that are used for hardware acceleration.

Two uvm_accel pipe proxy interfaces are required for the yamp example since bidirectional communication is required. Each uvm_accel pipe proxy interface is unidirectional; hence, the need for one input interface and one output interface. For most protocols, bidirectional communication is required. So, it is typical for two or more interfaces to be instantiated. Each uvm_accel pipe proxy interface takes a data item type as a parameter.

Standard UVM *methods* must be defined for each driver. These methods are customized using extensions depending on the abstraction level. In the yamp example the drive_transfer() method implements the driver functionality. Separating the code into distinct abstraction levels makes the code easier to understand and debug.

5.6.2.1 drive_transfer Method

When the master_bfm is extended to operate in UVM_ACCEL mode, acceleratable interfaces are used to send and receive data items as transactions from the HVL partition into the HDL partition where a HDL BFM drives the DUT signals. The drive_transfer() method does not implement any signal level protocol functionality; it operates purely at the transaction level in this mode. The HDL BFM is implemented as a separate module and is instantiated in the HDL partition that will be described in the next section.

The `uvm_accel` ports use standard transaction-level modeling (TLM) semantics to send and receive transactions via SCE-MI communication channels. The blocking `put()` and blocking `get()` functions are shown in the code snippet from the *e* yamp example below.

```
drive_transfer (cur_transfer : transfer_s)  @p_sys_smp.clk is only {
        cur_transfer.start_transfer();             // Get item from
                                                   // sequencer
        if (cur_transfer.direction == WRITE) {
            m_in$.put(cur_transfer);               // Drive write
                                                   // transaction
        }
        else if(cur_transfer.direction == READ) {
            var ref_data : list of uint (bits : YAMP_DATA_WIDTH) =
            cur_transfer.get_data().copy();
            cur_transfer.data.resize(0);           // reset the
                                                   // data
            m_in$.put(cur_transfer);               // Drive read
                                                   // transaction
            m_out$.get(cur_transfer);              // Get read data
        };
        cur_transfer.end_transfer();               // End current
                                                   // sequence
};
```

Once a data item has been taken from the sequencer it can be put into an input channel using the blocking `put()` function associated with the port that is bound to that channel. The blocking `put()` function blocks until the transaction has been taken from the channel at the opposite end. This means that the `drive_transfer()` method does not need to implement any sort of wait before informing the sequencer that the current sequence item has been done. This is simpler than in the non-accelerated case where you must implement any code required to allow one sequence to be completed before the next one is started.

5.6.2.2 Acceleratable Driver BFM (*e*)

The acceleratable driver BFM resides in the HDL partition and implements the signal level protocol functionality required to drive the DUT. The acceleratable driver BFM contains SCE-MI pipes interfaces which are bound to ports within the driver component which resides in the HVL partition. The `drive_transfer()` method passes transactions through a SCE-MI pipe to the driver BFM that must extract the transaction and apply it to the DUT signal level interface.

The acceleratable driver BFM must be written in acceleratable SystemVerilog or Verilog in order for it to be accelerated by a hardware accelerator. The driver code should be partitioned into separate files to distinguish between the code that is to be simulated and the code that is to be accelerated. This simplifies the overall compilation process and makes the code easier to maintain.

The UVM Acceleration package provides *e* and SystemVerilog interfaces to allow access to the pipes on the HVL side. The same acceleratable driver BFM can be used in both environments.

For more information about the acceleratable driver BFM, see Section 5.5.2.5, "Acceleratable Driver BFM (SystemVerilog)," on page 213.

5.7 Collector and Monitor

The collector and monitor components have a similar implementation to the driver and sequencer components described in the previous sections, except that the collector and monitor observe and track activity on the DUT interface rather than drive it.

The collector component is responsible for making the physical connection to the DUT and should use `abstraction_level` to determine the kind of interface that should be built during the UVM `build_phase` simulation phase in a similar fashion as previously described for the driver.

The main difference between a collector and a driver is that a collector is a passive component. It does not drive values onto the DUT interface. Therefore, it does not need to be impacted by the tri-stating of any of the signals. Apart from this, a collector should be architected and partitioned in a similar fashion to a driver.

5.8 Summary

Simulation performance can slow down to unacceptable levels when scaling the verification run to the chip or system level. Yet, the demand keeps rising to run such simulations to establish a higher level of confidence in the quality of the product being verified. The acceleratable Universal Verification Methodology (UVM) allows portions of a standard UVM environment to be accelerated using a hardware accelerator. In fact, the methodology does not restrict its usage to hardware acceleration alone. UVM acceleration is truly an extension of the standard simulation-only UVM, and is fully backwards compatible with it. This means that Universal Verification Components (UVCs) architected to be acceleratable can be used in either a simulation-only environment or a hardware-accelerated environment.

This chapter shows how UVM users can build acceleratable UVCs in either SystemVerilog or *e*. It describes how the UVC agent can be architected to operate in simulation as well as hardware acceleration. The underlying technology is compliant with the Accellera SCE-MI (Standard Co-Emulation API: Modeling Interface) standard providing additional vendor neutrality to the UVM community. In addition, the methodology is compliant with advanced verification techniques such as metric-driven verification, allowing the user community to further build additional verification intelligence into their verification arsenal.

6 Summary

The electronics industry grows by satiating a thirst for innovative features, performance, and connectivity with new and unique products. In an endless race for the leading edge, product teams turned to digital design for its relative simplicity and built sophisticated verification environments to deliver quality products. Bigger digital designs demanded better verification and better verification enabled still larger designs. The Accellera UVM arose in this environment as a means for large, distributed project teams to coordinate projects that spill across multiple companies and across geographies in an awesome display of engineering prowess and rapid standardization.

Awesome for sure, except that it is not a digital world. While the pundits pontificate from every media source about the digital home, digital car, and connected world, engineers actually have to design and build it. And engineers know the world is far from digital. It is analog in, out, and increasingly in-between. It is simultaneously high-performance and low-power. It is a fusion of hardware and software. And it demands more complexity delivered in a shorter time without sacrificing quality. Wow, it certainly sounds like job security for verification engineers!

And that's the magic—engineering solutions for hard problems is exactly our strength. The decision is whether the architecture that was standardized in the Accellera UVM can be extended to these advanced topics. Extension is ideal because it leverages existing industry-wide knowledge and provides a clear, incremental path forward. By no accident, a foundation has been built with the Accellera UVM that is scalable horizontally across projects and vertically through system verification. The chapters of this book introduce solutions that do just that—leverage the UVM to offer answers to the advanced topics facing verification teams.

"Offer" is the key word. Each of the topics presented in this book is production proven, but not yet standardized. With that said, every topic is built on existing standards and presented with minimal references to commercial products. Any company can take the information here and build products around it. Any topic can be driven by the industry to be standardized. In fact, one of the topics—extending the UVM for multiple languages—was just voted as a goal for the Accellera standards group, the Verification IP Technical Subcommittee, as this book was going to print.

The pundits may say it's a digital world. Let them. We're verification engineers. We know better and we are ready with solutions to prove it.

—Adam Sherer, Cadence

The Authors

Bishnupriya Bhattacharya – Architect, Cadence Design Systems

Author for the SystemC section of multi-language verification chapter.

Bishnupriya Bhattacharya has 14 years of experience in the EDA domain. She is an industry expert on the SystemC language and its various applications in design, verification, and in the ESL space. She has been actively involved in the SystemC language standardization process as it has evolved over the years. At Cadence, Bishnupriya is an R&D Architect for SystemC tools. She has a B.S. in Computer Science and Engineering from Jadavpur University, India, and a MS in Electrical Engineering from University of Maryland, College Park, USA.

John Decker– Solutions Architect, Cadence Design Systems

Co-author for the low-power verification chapter.

John Decker is a solutions architect for low power across the design and verification flow at Cadence. John has over 20 years experience in ASIC design and EDA, having held positions in design, application engineering, consulting, and R&D. John has several patents in the area of low-power verification and has been a key technical driver for the CPF language and LP methodology at Cadence for the last 5 years. He is the primary author of the Cadence Low-Power Methodology Guide and continues to work closely with key ASIC design companies to provide solutions for leading-edge low-power architectures and design techniques. John graduated with honors from Rensselaer Polytechnic Institute with a B.S. in Electrical and Computer Systems Engineering.

Gary Hall– Staff Solutions Engineer, Cadence Design Systems

Author for the acceleration chapter.

Gary Hall has over 15 years experience in ASIC and FPGA hardware design and verification spanning many sectors including defense, telecommunications and consumer. Gary graduated from The University of Portsmouth in 1995 with a B.E. with honors in Electrical and Electronic Engineering and has held engineering positions at GEC Plessey Semiconductors, Matsushita Telecommunications (Panasonic), Motorola and Cadence Design Systems.

Gary currently works in the Cadence Research & Development organization as a Staff Solutions Engineer for hardware emulation and acceleration solutions with special responsibility for the UVM Acceleration verification methodology.

Nick Heaton– Sr. Solutions Architect, Cadence Design Systems

Author for the metric-driven verification chapter.

Nick Heaton is an ASIC and EDA veteran with more than 25 years of experience in the design and verification of complex SoCs. Nick graduated from Brunel University, London in 1983 with First Class Honors in Engineering and Management Systems, initially working as an ASIC designer for ICL in Bracknell. In 1993, he founded specialist ASIC Design and Verification Company Excel Consultants, servicing customers such as ARM® and Altera. In 2002, Nick joined Verisity as Manager of Northern European Consulting Engineering.

Nick currently works in the Cadence Research & Development organization as a Senior Solutions Architect with special responsibility for the Incisive Verification Kit, a complex and realistic golden example of verification methodologies and technologies across all of Cadence front-end products.

Yaron Kashai– Distinguished Engineer, Cadence Design Systems

Co-author of the metric-driven verification of mixed-signal designs chapter.

Yaron has 24 years of EDA experience, mostly in design verification. His current focus is verification of digital and mixed-signal designs. Until recently Yaron served as VP R&D for new technology incubation at Cadence. He joined Cadence as part of the Verisity acquisition in 2005. Yaron was part of the Verisity founding team starting in 1996, where he served as vice president of research. Yaron began his career at National Semiconductor in 1985, working on distributed computing, and later managing the validation and test automation team.

Yaron has a B.S.E.E. cum laude from the Technion and MSEE from Tel Aviv University.

Neyaz Khan– Senior Scientist, Maxim Integrated Products

Co-author of the metric-driven verification of mixed-signal designs chapter.

Neyaz Khan is a Senior Scientist at Maxim Integrated Products. He has co-authored books on Low Power and Functional Verification: *A Practical Guide to Low-Power Design—User Experience with CPF–Si2*, and *Advanced Verification Techniques: A SystemC Based Approach For Successful Tapeout*, from Kluwer Academic Publishers. Neyaz holds a Masters degree in Electrical Engineering from Concordia University Montreal, Canada and a Bachelor of Engineering degree in Electronics & Telecommunications from REC/NIT Srinagar, India. Neyaz has worked for over 25yrs as design & verification lead at several companies: Cadence Design Systems, Texas Instruments, Bell Northern Research, CAE Electronics, Semiconductor Complex India. His areas of interest include low-power and mixed-signal design and verification.

He is involved with the development and deployment of Mixed-Signal and Low-Power Design & Verification products and methodologies at Cadence. Neyaz has played a key role in the deployment of the Common Power Format (CPF) for Low Power and the Incisive Verification platform at Cadence and has also co-authored books on Advanced Verification using SystemC. Neyaz has over 20 years of experience in ASIC Design and Verification. Prior to working for Cadence, he has served in technical-lead roles for a number of companies including Texas Instruments and Bell Northern Research.

Zeev Kirshenbaum – Solutions Architect, Cadence Design Systems

Author for the multi-language verification chapter.

Zeev has more than 14 years of experience during which he served in various roles at Verisity/Cadence including R&D, verification methodology development, and product engineering. Through these roles, Zeev has been involved with many different aspects of functional verification, both from the user-view and methodology perspective, as well as the in-depth technology implementation under the hood. For the past several years, Zeev has been focusing on developing methodologies to enable the use of OVM/UVM in complex multi-language system-oriented verification environments.

Zeev holds a B.Sc. cum laude in Computer Science from the Technion, Israeli Institute of Technology.

Efrat Shneydor – Solutions Engineer, Cadence Design Systems

Co-author for the low-power verification chapter.

Efrat Shneydor has worked in the functional verification field since 1992, since graduating the Hebrew university in Jerusalem. Efrat worked as a verification engineer in Digital Semi-Conductors, one of the pioneer users of Specman. Efrat joined Verisity in 1998, and was a co-founder of the Verisity methodology team, the group who developed the first verification reuse methodology—*e*RM. When Verisity was acquired by Cadence, Efrat joined Cadence methodology team, developing OVM and UVM. Efrat owns the development of UVM *e*, including module-to-system reuse, UVM *e* acceleration, low power, and more.

Contributors

Frank Armbruster, Sharon Rosenberg, Adam Sherer

Acknowledgements

Tom Anderson, Dave DeYoreo, Kristin Lietzke, Joseph Pizzi, Joern Stohmann, Amy Witherow

Index

A

acceleratable
    driver (*e*) 216
    driver (SV) 209
    driver BFM (*e*) 218
    driver BFM (SV) 213
    UVCs 197
        building in *e* 215
        building in SV 207
acceleration 195
    hardware 197
    interfaces 201
    packages 201
active bias 111
active_passive 75, 93
agent
    active 196
    passive 197
analog
    checking functionality 38
    input sources 71
    model creation 53
    properties, capturing 21
    regression runs 65
    verification, planning 20
architecture
    acceleratable UVC 198
    low power 105
    multi-language 155
    testbench 6
assertions 49
    advantage of 71
    low power 135
    power-aware 124

automated verification plan 136

B

build_phase function 210

C

checks
    low power equivalency 129
    power control sequences, automatic 135
    versus functional verification 130
clk_period 75, 88, 92
clocks 51
    driving 51
    gating 107
collector 219
configuration
    e info to SV 158
    parameters
        ramp UVC 92
        threshold monitor UVC 88
    recommendations for SystemVerilog adapter 148
    SV config containers 157
connecting
    e and SystemVerilog 147
    using TLM ports and SystemC 182
controls, power 51
coverage
    adding 70
    analog 63
    collecting 26
    direct and computed 27
    low power 135

ranges 30
timing collection 27

D

data items 207
data items (*e*) 215
debugging, UVCs 146
delay 75
directory structure 62
dms_reg 24
dms_register sequence item parameters 80
dms_threshold 25
dms_threshold_measurement 88
dms_wire 23
domains
    dependencies 114
    interfaces 139
    secondary 106
drive_seq 76
drive_transfer method (*e*) 217
drivers
    proxy, SystemC 192
    SystemC 191
duration 75
dynamic voltage 110

E

e
    and SystemVerilog, connecting 147
    calling methods from SystemVerilog 150
emulation, low power 130, 131
eRM and UVCs 146
events, passing across languages 153
example, UART module UVC 155

F

frequency response, checking 47

G

get_and_drive_accel() task 212
get_and_drive() task 212

H

hardware acceleration 195, 197
hardware interface, SCE-MI 205
has_coverage 75, 88, 93
HDL partition 198
hdl_path() 75, 88, 92
HVL and HDL partitions 197

I

Incisive Verification Kit workshops 14
inputs, generating 32
issues, low power, common 138

L

low power 8
    architecture 105
    assertions and coverage 135
    browser display of power state 132
    challenges 98
    common issues 138
    design and equivalency checking 128
    optimization 104
    resources 111
    sequence driver 126
    simulation and emulation 130
    structural and functional checks 129
    system-level control 140
    UVC 122, 124
    verification 97
        challenges 102, 103
        methodology 100
        planning 113
        scope 99
        tasks 121
    verification methodology 111
    visualizing influence on design 133
    waveform display of power state 132

M

measure_seq 76
metric-driven verification (MDV) 1, 15
min_ramp_time 92
min_ramp_voltage 92
modeling

configuration components, SystemC 175
SystemC 175
models
    creation, analog 53
    styles 57
modes
    monitor 75
    power 135
        recommendations 119
    standby 110
monitor 219
monitor mode 75
monitor_min_time 88
monitor_min_voltage 88
monitor_noise_threshold 92
monitor, using
    power-aware assertions and scoreboards 124
    system-level checking 124
multi-language UVM, overview 143
multiple supply voltages (MSV) 107

P

parameters
    ramp UVC configuration 92
    system-level for mixed-signal 56
    wire UVC configuration 75
partition components, hardware and software 198
partitions for hardware acceleration 197
planning
    domain-level verification 118
    hierarchical 117
    low-power verification 113
    power management 115
    system-level, for low power 113
port_cross 88
port_h_l_cross 88
port_l_h_cross 88
ports, connecting in SystemC 193
power
    consumption, reducing, table 104
    control sequences, checking 135
    controls 51
    domains 105
    intent, modeling 130
    management 115
    mode transitions 106

    modes 106
        legal 135
    shutoff 107
        verification considerations 109
power state, display in browser 132
project directory, recommended structure 62

R

regressions, multiple 67
requirements, UVC 146
resets 51
reuse 9
    IP verification elements 19
    verification plan 22
run_phase task 211

S

scalability 11
SCE-MI
    definition 200
    hardware interface 205
    input pipe interface 206
    output pipe interface 206
scoreboards, power-aware 124
sequence items
    for wire UVC 76
    parameters, dms_register UVC 80
sequences
    invoking SV from e 166
    and UVCs 146
simulation
    flow in mixed e and SystemVerilog 173
    low power 130
SoC level
    simulation 69
    test plan 70
    testbench 70
stimuli, generating and injecting, in SV 160
SystemC
    bi-directional ports 189
    driver 191
    driving TLM model 182
    library
        modeling and verification 175
        new features 174

using 174
modeling configurable components 175
ports, connecting 193
proxy driver 192
reference model
    issues 189
    SystemVerilog VE 188
reusing verification components 191
selecting connection 188
using TLM to connect languages 182
verifying using TLM ports 177
system-level
    checking using the monitor 124
    power behavior 117
SystemVerilog
    and *e*, connecting 147
    API over *e* 172
    calling from *e* 150
    over *e* 171
SystemVerilog UVC
    checking 168
    class-based system UVC, *e* over 154
    configuration 156
        config containers, defining 157
        config information from *e* 158
        SystemVerilog getting configuration information 159
    environment architecture 155
    generating stimuli 160
    injecting stimuli 160
    items, doing from *e* 161, 162, 166
    monitoring 168
    SystemVerilog sequences, invoking from *e* 166

T

test environment, integrating 54
testbench
    architecture
        details, kit 6
        simple for MS 25
    connecting 54
    integrating into SoC level 70
    model styles in one 57
    strong foundations 5
timing
    measuring 43

system-level for mixed-signal 56
TLM model, SystemC 182
TLM ports
    connecting between languages 182
    verifying, SystemC 177
tradeoff, speed and visibility 30
TRANS_RAMP_SEQ 93
type conversion 147

U

UART module UVC sample environment 155
unified verification components (UVCs) 143
UVCs
    acceleratable 197
    architecture 143
        acceleratable 198
    architecture diagram 145
    connecting from *e* to SV higher layers 171
    connecting the layers 147
    for simulation 196
    layers 144
    monitor 124
    overview of 143
    requirements for 146
UVM
    multi-language overview 143
    power-aware environment 121
uvm_accel_input_pipe_proxy (*e*) 204, 205
uvm_accel_input_pipe_proxy (SV) 203
uvm_accel_output_pipe_proxy (*e*) 205
uvm_accel_output_pipe_proxy (SV) 204
uvm_accel_pipe_proxy_base (*e*) 204
uvm_accel_pipe_proxy_base (SV) 202
UVM-MS 16
    architecture 26
    IP and SoC level 16
    VE, constructing 23

V

values, comparing to threshold 45
verification
    analog, planning 20
    components, reusing SystemC 191
    environment
        architecture for multi-language 155

    control register integration 79
    executing for low power 128
    foundations 5
    low power 97
        execution 128
    mixed signal 23
    power aware 121
    ramp UVC integration 90
    SC reference model 188
        example 188
    SoC level 19
    tasks for low power 121
    UVC integration steps 73
flow
    IP level 18
    SoC level 18
low power 97
low-power structures 119

plan 2, 17
    automated 136
        example 137
    metrics 5
    structured for reuse 22
planning 113
power aware 128
SystemC 175
voltage, dynamic 110
vPlan
    analog 63
    implementation data 64

W

workshops
    Incisive Verification Kit 14